SpringerBriefs in History of Science and Technology

More information about this series at http://www.springer.com/series/10085

Zdeňka Jastrzembská · Dagmar Pichová ·
Jan Zouhar

Czech Women Philosophers and Scientists

Springer

Zdeňka Jastrzembská
Department of Philosophy
Faculty of Arts
Masaryk University
Brno, Czech Republic

Dagmar Pichová
Department of Philosophy
Faculty of Arts
Masaryk University
Brno, Czech Republic

Jan Zouhar
Department of Philosophy
Faculty of Arts
Masaryk University
Brno, Czech Republic

ISSN 2211-4564 ISSN 2211-4572 (electronic)
SpringerBriefs in History of Science and Technology
ISBN 978-3-030-61515-4 ISBN 978-3-030-61516-1 (eBook)
https://doi.org/10.1007/978-3-030-61516-1

This Springer imprint is published by the registered company Springer Nature Switzerland AG
The registered company address is: Gewerbestrasse 11, 6330 Cham, Switzerland

Acknowledgements

We owe a debt of gratitude to Ruth Hagengruber from the Center for the History of Women Philosophers and Scientists at Paderborn University, who encouraged us to start the research on Czech women philosophers and scientists. Researchers at the Center for the History of Women Philosophers and Scientists—Ana Rodrigues, Stefanie Ertz, Julia Lerius, and Andrea Reichenberger—helped us discuss various aspects of the project during the Paderborn seminar on women philosophers and scientists in May 2018. The volume in preparation was presented and discussed at the Institute of Philosophy at the University of Wroclaw during the workshop organised by Agnieszka Bandura in November 2018 and at the 17th IAPh conference at Tsinghua University in Beijing in August 2018. We would like to thank Herta Nagl-Docekal (University of Vienna), Mary Ellen Waithe (University of Cleveland) and our colleagues from Masaryk University, particularly Radim Bělohrad, Radim Brázda, Jakub Mácha, Denisa Nečasová, Helena Pavlincová, Marek Picha, and Daniel Špelda, for their helpful comments on the project.

The completion of the manuscript was possible thanks to the efforts of our translator Petra Trávníková and proofreader Anne Johnson. We would like to also thank Jana Svobodová and Danica Ježová for their help with the bibliography, name index, and manuscript editing.

The research was financially supported by the Czech Science Foundation, within the project 'Czech Women Philosophers and Scientists' (GA17-06697S).

Contents

About the Authors

Zdeňka Jastrzembská is an Assistant Professor at the Department of Philosophy, Masaryk University in Brno (Czech Republic). She specialises in epistemology, methodology and philosophy of science, with a particular emphasis upon theories of causation and explanation. Her publications include *Kauzální aspekty vysvětlení* [*Causal Aspects of Explanation*] (2007), *Aspekty vysvětlení: hledání explanačních znalostí* [*Aspects of Explanation: Searching for Explanatory Knowledge*] (2009), 'Perspektivistická genealogie kauzality' ['The Perspectival Genealogy of Causation'] *Organon F*, Vol. 17, No. 4 (2010), and 'Spor o vivisekce a české ženy na přelomu 19. a 20. století' ['The Controversy over Vivisection and Czech Women at the Turn of the twentieth Century'] *Teorie vědy*, Vol. 39, No. 2 (2017).

Dagmar Pichová is an Associate Professor at the Department of Philosophy, Masaryk University in Brno (Czech Republic). In her scientific and research activities, she focuses on the history of women philosophers and 18th century French philosophy and literature. She is the author of *Émilie Du Châtelet, femme de lettres* (2018), 'Émilie Du Châtelet: devenir femme de lettres.' In *La condition des femmes dans l'Europe du XVIIIe siècle*. Lumières, n° 24 (2015), *100 myšlenkových experimentů ve filozofii* [*100 Thought Experiments in Philosophy*] (with Marek Picha, 2013), and *La communication ironique dans Le Roman comique de Paul Scarron. Étude comparative avec Don Quichotte de Cervantès* (2007).

Jan Zouhar is a Professor at the Department of Philosophy, Masaryk University in Brno (Czech Republic). An eminent specialist on Czech philosophy, he has published many books and articles devoted to the discipline, e.g., *Česká filosofie v letech 1945–1948* [*Czech Philosophy 1945–1948*] (with Helena Pavlincová and Jiří Gabriel, 2013), *Česká filozofie v šedesátých letech* [*Czech Philosophy in the 1960s*] (2009), *O Masarykovi* [*On Masaryk*] (2009), *Dějiny českého filozofického myšlení do roku 1968* [*History of Czech Philosophical Thinking to 1968*] (2008), *Česká filosofie v letech protektorátu* [*Czech Philosophy during the Protectorate*] (with Helena Pavlincová and Jiří Gabriel, 2007), and 'On Contemporary Czech Philosophy' In *Philosophy Worldwide: Current Situation* (2007).

Chapter 1
Introduction

The last few decades have witnessed an increasing interest in the history of women philosophers and scientists in various cultural contexts. Since the pioneering edition of *A History of Women Philosophers* (1987–1995) by Mary Ellen Waithe, historical research has shone a light on many women whose work was unjustly disregarded. It is our aim to contribute to the research completing the gaps in the traditional history of philosophy and science by turning attention to the women in the Czech territory.

This volume presents an overview of the most important Czech women philosophers and scientists, focusing on the context of their lives and work. We stress the complex conditions of women in Czech history in relation to the position of learned women in other European countries. The main goal is to present a systematic description of the impact of women on the history of Czech philosophy and science, based on the analyses of their works and the characteristics of contemporary debates in which they participated.

1.1 Methodology

Before we present a chronological overview, it is necessary to introduce the main axes of the research and describe the methodology applied. Although our main goal is the work and activities of women in the field of philosophy, we aim to study women in both philosophy and science because of the common history of these two approaches to the world and its interpretation; after all, the modern notion of science has been in use only since the 1700s. We present the activities of Czech women in a variety of academic fields, including humanities, natural sciences, and technical disciplines.

Z. Jastrzembská et al., *Czech Women Philosophers and Scientists*,
SpringerBriefs in History of Science and Technology,
https://doi.org/10.1007/978-3-030-61516-1_1

A project combining philosophy and science can also be justified by current research in the historiography of science: approximately since the mid-1960s, a number of methodological principles have been formulated that can also be applied to the history of philosophy. *An Introduction to the Historiography of Science* (Kragh 1987) discusses some methodological issues which are of fundamental importance to our research and which we considered when studying Czech women philosophers and scientists. In this text, we observe the three most important principles, which we identify as sources, relevance, and context:

Sources—traditionally, the position of a philosopher or a scientist in the history of the discipline is related to the books they authored and published. The study of correspondence, diaries, translations, autobiographies, memoirs, etc. was viewed as secondary. Following the modern historiography of science, we consider that to determine the history of a particular discipline, the ideas expressed in diaries or commentaries to translations are relevant as well, even if they lack systematic presentation. This principle helps to include women in the history of philosophy and science, especially in the earliest periods when women's publishing activity in the Czech territory was meagre.

Relevance—women are still sometimes excluded from the history of philosophy and science because their ideas are not considered to be relevant, based on the present evaluation of ideas or on particular approaches to philosophy or science, or judged from the point of view of contemporary reception and discussions. The figures of women philosophers and scientists can be studied by adopting an approach exempt from prejudice, i.e. research without any particular conceptions of what is relevant or crucial in the disciplines.

Context—one of the main contributions of the modern historiography of science lies in stressing the importance of the broader intellectual, cultural, and social context of the formation of science. This principle is also crucial for our text. For example, a detailed study of social connections can show that even if women were not able to fully participate in the research and formation of knowledge without special education, it is possible to prove that they were interested in scientific and philosophical questions, formulated in the context of personal contemplation or through practical activities.

We situate our book within the primary research in the field of the historiography of science and history of philosophy. Even though we are also motivated by the research done in the field of feminist philosophy and gender studies, we do not follow their methods and terminology. We hope, however, that the outcome of our research could be a source of inspiration for future studies, developing the themes outlined in the text from the perspective of other disciplines and methodological schemes.

In research devoted to Czech women in science and philosophy up until the 1800s, restricting the selection of women to only those writing in Czech would mean a great impoverishment. Hence the choice of individual figures will be determined by their place of origin—it will concern women born on the present-day territory of the Czech Republic or coming from traditional Bohemian and Moravian aristocracies. The body of work is extended by authors who did not work solely within this territory but also spent some time abroad without losing touch with the Czech environment. This extension applies especially to the 1900s, marked by several waves of emigration. We are aware of the close connections between Czech and Slovak intellectual activities; however, we do not study the situation in Slovakia or related to scholars working in Slovak institutions (cf. e.g. Ďuranová et al. 2003; Kodajová et al. 2019).

Even though women in the Czech territory rarely published their texts before the 1800s, there is still proof of their writing activities. In some cases, manuscripts of their texts are available. Their correspondence, their diaries, and the selections of literature they purchased for their private libraries are important evidence of women's interests. The preserved texts written by Czech women are mostly written in French and German and can be accessed in archives and chateau libraries.

In our approach to preserved documents, we stress the philosophical and scientific interests of selected Czech women; we do not aim to provide detailed information about their lives. Our intention is to connect the philosophical and scientific interests of Czech women with the contemporary European context. At the same time, we would like to show the presented documents in relation to present-day philosophical and scientific discussions, translating their intentions and conceptual apparatus into modern philosophical and scientific terminology.

1.2 Bibliography

The field of history of women philosophers and scientists started to be explored in the last decades of the 1900s. Starting in the 1980s, it is possible to observe a rise of interest in the history of philosophy in the publications of feminist philosophers. The main outcomes of the research can be seen in feminist (re)interpretations of the work of important individuals in the history of philosophy (e.g. Bar On 1994; Mills 1996) and in the rediscovery of women philosophers (e.g. Tuana 1992; Atherton 1994; Warnock 1996). The representatives of feminist philosophy have emphasised the overly misogynistic character of traditional philosophy, proposed a gender-based interpretation of key philosophical notions, and offered their own conception of rationality (cf. e.g. Bordo 1987; Lloyd 2002; Alanen and Witt 2004; Kourany 1998). Thanks to feminist criticism, a serious discussion on the role of women and the research of the missing chapters in the traditional conception of the history of philosophy and science has been initiated.

During the feminist polemics with traditional philosophy, the following questions have been raised: What are the principles behind the formation of the philosophical canon? Which criteria must the authors and works fulfil in order to become a part? Is the representation of philosophy without women typical of all historical periods? How could women be included again in the history of philosophy? (cf. e.g. Witt 2006) Mostly due to the interest in feminist philosophy, women started to appear in prestigious philosophical encyclopaedias and philosophical and historical overviews, culminating partly in the aforementioned four volumes of *A History of Women Philosophers* (Waithe 1987–1995), which deals with women philosophers and scientists in the period from antiquity to the present. Recent interest in the theme can also be observed in the monothematic issue of the *Monist* philosophical journal (2015) and the *Women in the History of Philosophy and Sciences* Springer Series edited by Ruth Hagengruber, Mary Ellen Waithe, and Gianenrico Paganini.

In Czech research, women are a particular object of interest in the field of history within the scope of women's history. Works deal with the status of women in society from the medieval period to the first half of the 1900s (cf. e.g. Pešek and Ledvinka 1996; Lenderová 2002, 2009; Čadková et al. 2006). New approaches were proposed from the perspective of gender studies (Storchová and Ratajová 2013).

The emancipatory movement, starting in the 1800s, is also well described (e.g. Neudorflová 1999; Bahenská 2005; Bahenská et al. 2011). Attention is paid to women in the history of literature and art; research has recently focused on gender and sociological studies mapping the professional and personal lives of women in science (e.g. Štrbáňová et al. 2004; Havelková and Oates-Indruchová 2014; Wagnerová 2017, and the activities of the Centre for Gender and Science).

The situation is different in the history of Czech philosophy. The *Slovník českých filozofů* [*Dictionary of Czech Philosophers*] (Gabriel et al. 1998) offers a basic orientation in the lives and work of Czech women philosophers; their names can also be found in various works about the history of Czech philosophy (cf. e.g. books by Zouhar et al. 2003, 2006, 2007, 2013; Gabriel et al. 1994; Hoffmannová 2016) and their texts appear in anthologies covering discussions and controversies in Czech philosophy (cf. e.g. Zouhar et al. 2003). However, there is still no systematic publication in the Czech language that presents, analyses, and evaluates the work of Czech women philosophers and scientists. The books written in Czech discuss various topics related to women in science and philosophy; however, a general overview of the theme is still missing.

To provide a clear introduction for readers, this volume respects a chronological order, from the earliest period to 1989, the year of the 'Velvet Revolution' that marked the end of socialist Czechoslovakia. The chapters are completed by a chronological timeline that summarises Czech history and situates the presented women within a historical context.

Due to the limited number of preserved documents, the first chapter devoted to the period before the 1820s (*Before the 1820s: Setting the Stage for Women's Intellectual Progress*) is rather brief compared to the two subsequent chapters. Nevertheless, we hope that further research will explore the broad range of the philosophical and scientific interests of Czech women before the 1800s, based on an analysis of preserved books and manuscripts (e.g. the translations by Maria Eleanora and Anna Katharina von Sporck, and the moral and pedagogical writings by Alexandra Shuvalova).

The central chapter (*1820s–1918: An Expansion of Women's Intellectual Pursuits*), describing Czech women's activities in the 1800s and early 1900s (until 1918, the foundation of the independent state of Czechoslovakia), opens various perspectives for future research, as there are many documents (correspondences, publications, manuscripts) that merit further study and detailed commentary.

In the volume, the main focus is concentrated on the field of philosophy, especially in the 1900s—the last chapter, covering the period 1918–1989 (*Czech Women in

Philosophy and Science since 1918), is selective and refers only to the most important Czech women scientists and philosophers. The overview offers various research questions and new motivations for study in the field of history of women philosophers and scientists.

We present a typology of women philosophers and scientists based on various aspects in each chapter. The lives and works of these women serve as case studies that differ in the historical periods we describe, showing the ways Czech women approached philosophical and scientific themes and gradually entered the academic institutions of Czech science.

Zdeňka Jastrzembská, Dagmar Pichová, Jan Zouhar
Brno, Czech Republic
September 2019

References

Alanen L, Witt C (eds) (2004) Feminist reflections on the history of philosophy. Kluwer Academic Publishers, Dordrecht/Boston

Atherton M (1994) Women philosophers of the early modern period. Hackett Publishing Co., Indianapolis

Bahenská M (2005) Počátky emancipace žen v Čechách: dívčí vzdělávání a ženské spolky v Praze v 19. století. Libri, Praha

Bahenská M, Musilová D, Heczková L (2011) Iluze spásy: české feministické myšlení 19. a 20. století. Veduta, České Budějovice

Bar On B-A (ed) (1994) Engendering origins: critical feminist readings in Plato and Aristotle. State University of New York Press, Albany

Bordo S (1987) The flight to objectivity: essays on Cartesianism and culture. State University of New York Press, Albany

Čadková K, Lenderová M, Stráníková J (eds) (2006) Dějiny žen, aneb, Evropská žena od středověku do poloviny 20. století v zajetí historiografie. Univerzita Pardubice, Pardubice

Ďuranová Ľ, Šourková A, Táborecká A (eds) (2003) Lexikón slovenských žien. Slovenská národná knižnica, Martin

Gabriel J et al (eds) (1998) Slovník českých filozofů. Masarykova univerzita, Brno

Gabriel J, Hroch J, Nový L (eds) (1994) Czech philosophy in the XXth century. Council for research in values and philosophy, Washington

Havelková H, Oates-Indruchová L (eds) (2014) The politics of gender culture under state socialism: an expropriated voice. Routledge, London

Hoffmannová J (2016) Prvenství žen. Ženy iniciativní, vzdělané a tvořivé. Ústav T. G. Masaryka, Praha

Kodajová D et al (2019) Živena. 150 rokov spolku slovenských žien. Slovart, Bratislava

Kourany JA (ed) (1998) Philosophy in a feminist voice: critiques and reconstructions. Princeton University Press, Princeton

Kragh H (1987) An introduction to the historiography of science. Cambridge University Press, Cambridge

Lenderová M (ed) (2002) Eva nejen v ráji. Žena v Čechách od středověku do 19. století. Karolinum, Praha

Lenderová M (ed) (2009) Žena v českých zemích od středověku do 20. století. Nakladatelství Lidové noviny, Praha

Lloyd G (ed) (2002) Feminism and history of philosophy. Oxford University Press, Oxford

Mills PJ (ed) (1996) Feminist interpretations of G.W.F. Hegel. The Pennsylvania State University Press, University Park, PA
Monist (2015) The history of women's ideas 98(1)
Neudorflová ML (1999) České ženy v 19. století: úsilí a sny, úspěchy i zklamání na cestě k emancipaci. Janua, Praha
Pešek J, Ledvinka V (eds) (1996) Žena v dějinách Prahy. Scriptorium, Praha
Storchová L, Ratajová J (2013) Gender a bádání o raném novověku. In Koldinská Šedivá M, Cerman I (eds) Základní problémy studia raného novověku. Nakladatelství Lidové noviny, Praha
Štrbáňová S, Stamhuis IH, Mojsejová K (eds) (2004) Women scholars and institutions, proceedings of the international conference, Prague, 8–11 June 2003. Studies in the history of sciences and humanities, 13 A. Výzkumné centrum pro dějiny vědy, Praha
Tuana N (1992) Woman and the history of philosophy. Paragon Press, New York
Wagnerová A (2017) Žena za socialismu. Československo 1945–1974 a reflexe vývoje před rokem 1989 a po něm. Slon, Praha
Waithe ME (ed) (1987) A history of women philosophers: ancient women philosophers 600 B.C.–500 A.D. Kluwer Academic Publishing, Dordrecht
Waithe ME (ed) (1989) A history of women philosophers: medieval, renaissance and enlightenment women philosophers A.D. 500–1600. Kluwer Academic Publishing, Dordrecht
Waithe ME (ed) (1991) A history of women philosophers: modern women philosophers, 1600–1900. Kluwer Academic Publishing, Dordrecht
Waithe ME (ed) (1995) A history of women philosophers: contemporary women philosophers, 1900–Today. Kluwer Academic Publishing, Dordrecht
Warnock M (ed) (1996) Women philosophers. J.M. Dent, London
Witt C (2006) Feminist Interpretations of the philosophical canon. Signs: J Women Cult Soc 31:537–552
Zouhar J, Pavlincová H, Gabriel J (2003) Spory v české filosofii mezi dvěma světovými válkami: výbor textů. Masarykova univerzita, Brno
Zouhar J, Pavlincová H, Gabriel J (2006) Filosofie za protektorátu: vybrané texty z českých filosofických časopisů. Masarykova univerzita, Brno
Zouhar J, Pavlincová H, Gabriel J (2007) Česká filosofie v letech protektorátu: poznámky k tématu. Masarykova univerzita, Brno
Zouhar J, Pavlincová H, Gabriel J (2013) Česká filosofie v letech 1945–1948. Academicus, Brno
The Centre for Gender and Science (2019) http://genderaveda.cz/en/gender-and-science/. Accessed 14 Jan 2019
International project Women's Memory (2003) Gender studies. http://www.womensmemory.net. Accessed 14 Jan 2019
Ženy v disentu (2019) Sociologický ústav AV ČR. http://zenyvdisentu.soc.cas.cz/literatura. Accessed 14 Jan 2019
Ženy ve vědě do roku 1945 (2013) 3.0 Česko. http://albina.ff.cuni.cz/index.php/Hlavn%C3%AD_strana. Accessed 14 Jan 2019

Chapter 2
Before the 1820s: Setting the Stage for Women's Intellectual Progress

Abstract Most women in the earliest periods of Czech history had access to education only in convents. The Renaissance period saw educated women at the court of Rudolf II, and women with philosophical and scientific interests could be found in the Jewish community in Prague as well. Written documents, including argumentative texts intended to defend the church, prove women's intellectual interests in the religious reformation movements of the 15th and 16th centuries. Francophonie developed among the Czech nobility in the second half of the 1600s. The preserved manuscripts of women authors of that time were predominantly in French, and their translations were most commonly from French as well. The most educated women in the Czech territory included Maria Eleanora and Anna Katharina von Sporck, who translated religious and philosophical French literature. The philosophical and scientific interests of Czech women are also evident from the lists of books acquired by private chateau libraries, e.g., in Český Krumlov, initiated by Marie Ernestine von Eggenberg. A new genre that engaged aristocrats in the 1700s was educational and moral contemplations. This interest was related to the new emphasis on the role of the mother, stemming mainly from the ideas of Jean-Jacques Rousseau.

2.1 Earliest Period

At the beginning of recorded Czech history, there was a legendary ruler named Libuše (Libussa in English). She is mentioned by Cosmas of Prague in his chronicle written in Latin in the early 1100s, and the legend is further elaborated in the *Chronicle of Dalimil* in the 1300s (cf. Třeštík 1966). Libuše's exceptional character was noted by Cosmas, who described her as 'truly a woman among other women, cautious in counsel, quick to speak, chaste in body, upright in character, second to no one in resolving the lawsuits of the people. Affable, even lovable, in all things, she adorned and glorified the feminine sex while handling masculine affairs with foresight.' (*Kosmova kronika česká* 2005, 25) Even though Libuše eventually decided to

Z. Jastrzembská et al., *Czech Women Philosophers and Scientists*, SpringerBriefs in History of Science and Technology, https://doi.org/10.1007/978-3-030-61516-1_2

marry and handed over the rule to her husband Přemysl, the legend introduced in the Czech culture the image of a wise woman deciding for herself about her life and the fate of her country. The character of legendary Libuše was often referred to in the 19th century Czech national revival; Bedřich Smetana composed a patriotic opera dedicated to this subject, first performed in 1881 at the National Theatre in Prague.

Despite the legend of Libuše, the medieval approach to women and their intellectual abilities within the Czech territory, as in other European countries, stemmed from the Christian doctrine based on the theories of Aristotle and Galen. Women were subordinate to men and were to develop traditional womanly virtues related to childbearing and housekeeping. Nevertheless, the medieval Czech lands did not feature manifestations of strong misogyny; historians believe this may be due to the slower spread of Christianity and the subsequent development of Marian devotion (cf. Kopičková 2002, 19). This trend is also evidenced by the limited extent to which witch hunts took place in the Czech territory towards the close of the 1600s.

With some exceptions, often related to family background, women in the earliest periods had access to education only in convents or through movements associated with them. Hence those women's activities that can be regarded as manifestations of philosophical or scientific interest were closely connected to religious topics. The first female convent as well as the first monastery in Bohemia was the Benedictine Convent of St George, founded in Prague around 976 by Mlada (930/935–994), the sister of Boleslaus II. Mlada also managed to obtain consent from Rome to the establishment of a diocese in Prague, as noted by the aforementioned chronicler Cosmas in his text: 'He had a sister called Mlada, a virgin devoted to God, learned in sacred letters, given over to the Christian religion, gifted with humility, charming to talk to, a generous patron of paupers and orphans, and adorned with every kind of integrity of character. When she went to Rome for the sake of prayer, she was received benevolently by the pope. Living there for some time, she was imbued fully with monastic discipline. Finally, the lord pope, upon the advice of his cardinals and indeed wanting to help the new church by his beneficent judgement, consecrated her as abbess, having changed her name to Mary, and gave her the rule of St Benedict and the abbatial staff. Afterwards, having received permission and apostolic blessing to bring this new and holy monastic rule to Bohemia, the new abbess rode back to her sweet fatherland with her companions very delighted.' (*Kosmova kronika česká* 2005, 49).

Another prominent learned woman in the Czech convents was the Přemyslid princess Agnes (1211–1282), who was the abbess of the Poor Clares monastery 'Na Františku' in Prague. Agnes could speak several languages and exchanged letters with leading figures in the church. She drew inspiration from Francis and Clare of Assisi and was well known for her charity. Her sister Vilemína (Guglielma in English), who died in 1279, was a key figure as well; according to some scholars, she was the founder of the Milanese sect of Guglielmites. Her adherents pursued feminist aims—founding a female church (including a female pope and cardinals), which was to replace the male hierarchy of that time. However, Vilemína's remains were burnt at the stake in 1300 and her sect was declared to be heretical by the

Inquisition. There is a great deal of mystery surrounding her life even today (cf. Dinzelbacher 2003, 37, 1993; Muraro 1987).

At the beginning of the 1300s, the mystical inclination of learned Czech women (along the lines of Rhenish female convents and the Beguine movement) was continued by Agnes's niece Kunhuta (Kunigunde in English, 1265–1321), the abbess of St George's Convent (cf. Kopičková 2008, 18). Kunhuta contributed a great deal to enriching the convent library. She commissioned the local scriptorium to adapt older manuscripts and write new books. The most significant work was the *Passional*, which describes Kunhuta's mystical relationship to Christ. Kunhuta herself is the main heroine of the manuscript and Christ's 'bride', a frequent representation appearing in medieval women's mysticism. In the early 1300s, in 1307–1311, the German mystic Meister Eckhart von Hochheim lived in Bohemia. He advocated the notion of the *unio mystica*. For a short time, Meister Eckhart was the superior of the Dominicans from St Clement's convent, including the friar Kolda of Koldice, who authored most of the texts of Kunhuta's *Passional* (Kopičková 2008, 19–21). Nonetheless, scholars disagree about the degree of Eckhart's direct influence on Kolda's texts (cf. Sokol 1993, 31; Urbánková and Stejskal 1975, 29).

Kunhuta inspired her niece, Elizabeth of Bohemia (1292–1330), the mother of Charles IV, who dwelt with her in the monastery as a young girl. Elizabeth and her stepmother Elizabeth Richeza were prominent figures in Czech history, also supporting the arts and the production of religious books. One particularly remarkable cultural feat is the illuminated manuscripts commissioned by Elizabeth Richeza in the scriptorium in Brno and given to St Thomas's Abbey.

In addition to the contemporary evidence about women living in monasteries, there are mentions of educated women from aristocratic circles and their intellectual interests. For example, Charles IV was surrounded by educated women: his sister Jutta von Luxemburg (1315–1349) became a prominent patroness of the arts in France and his third wife Anna von Schweidnitz (1339–1362) corresponded with Petrarch. Unfortunately, her letters have not survived, although Petrarch's answer to Anna's announcement of giving birth to a daughter from 23 May 1358, in which the poet (presumably responding to her disappointment) congratulates her and mentions prominent women from the past to console her, is known (Špička 2010, 38). Anna von Schweidnitz grew up in the royal court of Buda, where she received an excellent education and upbringing under the influence of the progressive Hungarian queen Elizabeth.

Charles's daughter Anne (1366–1394), married to King Richard II of England, was also a distinguished person. She was well known for her intercessions on behalf of convicts, entering the cultural awareness as 'good queen Anne'. Anne brought with her to England precious illuminated manuscripts, including Czech and German translations of the Gospels. These translations of the Bible influenced similar attempts at translating in England led by John Wycliffe (cf. Kavka 2002, 174).

2.2 Renaissance and Reformation

The Renaissance period saw a rise in educated women in the Czech environment. Their extraordinary education most frequently had something to do with their family background; girls' intellectual interests were endorsed in educated families. Even though women's intellectual activities were often related to religion, they were not restricted only to convents.

The most extensive education in the period of humanism might have been received by Catharina Elisabeth von Kamenek (Elisabetha Albertina von Kammeneck). Catharina was a daughter of Nicholas Albert von Kamenek (c. 1550–1617), a prominent scholar who was a professor at Charles University and one of the translators of the Bible of Kralice. She knew, *inter alia*, Latin, Ancient Greek, and Hebrew. Her unusual education resulted from her family background; her brother was said to have been a competent translator and her sister Anna obtained a good education as well. Catharina is known to have written a poem in Latin and a letter to Balthasar Exner in which she commented on the education of girls, stating that it was not regarded with favour by her contemporaries (Stevenson 2005, 250).

Another outstanding educated woman of the same period was the poet Elizabeth Jane Weston (1581–1612), a stepdaughter of the alchemist and adventurer Edward Kelley. Kelley lived in the court of Rudolph II and his family accompanied him to Bohemia. Thanks to her stepfather's support, Weston was home-schooled by a tutor, receiving an excellent education, and she could speak several languages fluently. She stayed in Prague even after Kelley's death and appealed to influential figures in the imperial court with rhymed pleas to protect the family property. Gradually, she got involved in the epistolary communication of the Republic of Letters. During her life, she published two collections of poetry written in Latin: *Poemata* (1602) and *Parthenicon* (1610), which included a catalogue of learned women (cf. Westonie 2003; Storchová 2003). Weston gained the respect of the scholars of that period (her correspondents included the philologist Scaliger and the physician Croll) and was the only woman to earn her own entry in Pelcl's catalogue (Pelcl 1773–1782).

The sister of the astronomer Tycho Brahe, Sophie Brahe (1559–1643), also stayed briefly at the court of Rudolph II. She dealt with natural sciences (especially botany and chemistry) and assisted her brother with his astronomical observations (Christianson 2002, 30–45). Tycho thought highly of his sister's abilities and knowledge, as proven by his essay in which he explains why he intends to include his sister's letter into published scholarly correspondence.[1] Sophie herself carried out measurements at the royal observatory in Uraniborg (Uranienborg in Danish), which was built by her brother on the island of Hven. She came to Prague with her second husband, alchemist Erik Lange. Upon his death in 1613, she returned to Denmark. Unfortunately, with the exception of an extensive genealogical study, Sophie Brahe pursued her scholarly interests in private and did not publish the findings of her research.

[1] John Robert Christianson claims that Sophie's letter was not preserved and the prepared volume of scholarly correspondence was also not produced as Brahe died before it could be published (Christianson 2002, 32, 42).

The Jewish community in Prague was another circle comprising women with philosophical and scientific interests. The situation of Jews in the Czech lands was positive in the 1500s and 1600s, in comparison with Poland and the German states. Especially under the reign of Rudolph II, the cultural and intellectual life of the Prague Jewish community flourished (cf. e.g. Putík 2009).

Rebecca (Rivka) Bat Meir Tiktiner, who died in Prague around 1550, was probably of Polish origin. She wrote an ethical and medical treatise (*musar*) in Yiddish for women, called *Meneket Rivkah* [*Rebecca's Nursemaid*], which was published in Prague in 1609. Even though the author based it on the structure of a traditional religious treatise, she referred in her text to contemporary medical authorities. Rebecca had a thorough education based on the rabbinical tradition (both her father and husband might have been rabbis) (Hoffmannová 2016, 27; Tiktiner 2009). In his preface, the Prague publisher of her book expresses respect and admiration for the author's personality and works and explains his motivation: 'I had it printed so that every woman reading it will grasp the book and buy it and the title of the book is *Meneket Rivkah* to commemorate her and be to all women's credit that a woman can also write words of ethical instruction and good Biblical interpretation, just as good as many men…' (quoted acc. to Muneles and Vilímková 1955, 247–248; cf. Šedinová 1996, 91–100).

There were other educated women in the Prague Jewish community, including Bela Horwitz and Rachel Raudnitz, authors of the prosaic *Ein schön maase es ist geschehn ehe noch Jehudim haben zu Prag nicht gewolmt* [*Beautiful Story from the Period When the Jews Did Not Live in Prague*], published in Prague in the early 1700s. Chava (Eva) Bacharach (c. 1580–1651), a granddaughter of the legendary Rabbi Löw, was regarded as the most learned woman of her time. Sadly, Bacharach did not leave behind any written evidence, and the exceptional wisdom she used to raise her children and grandchildren is known only from the testimonies of her family members, especially her son Šimšon and her grandson Yair Chayim (Šedinová 1996, 99).

This period also included the deeds of the noblewoman Alena Meziříčská of Lomnice, who died in 1585; on her estate in Velké Meziříčí, she founded a Latin Lutheran academy (grammar school) in 1576. Unfortunately, the school operated only for two decades and was closed down after the estate changed owners in 1602 (Hoffmannová 2016, 25). Even though the institution was run only for a short time, it was attended by some prominent figures from Czech history, such as Jindřich Matyáš Thurn and Zdeněk Brtnický of Waldstein. The humanistic education in the Czech lands was shaped to a certain extent also by Anna Adam of Aventino (1550–1605), the daughter of Jiří Melantrich of Aventino, who built a large printing workshop. Anna, who was married to a major humanistic intellectual, Daniel Adam of Veleslavín, inherited the workshop and went on printing books and supporting the quality of education.

One of the best-known European women with philosophical ambitions, Elisabeth of the Palatinate (Princess of Bohemia) is related to the Czech lands only distantly, despite her name. Her mother Elisabeth Stuart was, together with her husband Frederick of the Palatinate, crowned a Czech ruler in 1619; in 1620, the royal family was forced to escape from Prague and lost the throne for good. The young Elisabeth, born in 1618, grew up in Brandenburg and The Hague and did not come into contact with the Czech environment.

In the religious reformation movements of the 15th and 16th centuries, Czech women became partially emancipated, even though their educational opportunities remained rather restricted. Women were already striving to be involved in theological disputes in the Hussite period (cf. Klassen 1999). Even those coming from lower classes were learning to read; some were said to know the Bible better than many priests. Nevertheless, most scholars and reformers of that time looked upon this activity on the part of women with disdain. Among the representatives of the Hussite movement (except for Matthew of Janow), traditional ideas of the role of women persevered.

Within the Czech reformation movements, women from the Unity of the Brethren were the most active participants in running the congregation and defending the movement. In the Unity of the Brethren, even girls were to obtain elementary educations. Further on, the congregation was organised so that older sisters were to take care of younger members and help them read and understand the Bible. Written documents reflecting the activities of women authors have survived from this period in the Unity of the Brethren. For example, Crescencia Zmrzlikova of Svojsin wrote a defence of the Unity of the Brethren in response to a letter of the Utraquist priest John Bechynka, and Johanna Kreiger of Krajek authored a written confession of faith. Moreover, Marta von Boskowitz (1463–c. 1509) prepared the defence of the Unity of the Brethren in a letter addressed to Vladislaus II in 1507 (Bránský 2008, 26–59). Marta's letter is a humble intercession, even though the author clearly stood up for her faith, remarking that the persecution of the Unity of the Brethren could arouse God's wrath: 'I was unhappy to hear that, especially for the sake of Your Grace, as I fear God's wrath and that Your Grace may incur it unwittingly […] And I do not know why I should deserve such condemnation, to be expelled from the country or put to death, because I would like to do everything that is virtuous. But I would never give up for anybody what I consider to be virtuous, unless I was shown or announced something better.' (Bránský 2008, 38) Even though the king did not comply with her request, his answer expressed his personal respect for the author, as is clear from his addressing her as 'Noble Marta von Boskowitz, dear to us'. Bohuslav Hasištejnský of Lobkowitz responded to Marta's letter in the opposite way; he prepared a stylistically refined polemic text in Latin, addressed to his friend Jan Pibra in 1508, in which he refers to stereotypical ideas of the role of women (Bránský 2008, 43–44).

Even though they are not elaborate theological contemplations, these written documents prove women's intellectual interest in argumentative texts determined to defend the church. Women from the Unity of the Brethren also followed their families on foreign missions. For example, the Moravian expatriate Rosina Zeisberger, née Schneider, lived in North America; Rosina Beck, née Stach, was the first missionary in Greenland; and Anna Maria Tonn lived in Suriname (Štěříková 2014, 271–364). However, documents proving their possible (scientific) interest in these exotic destinations are not available, other than documents confirming their efforts to learn the local languages.

There is contemporary evidence of intellectually ambitious women who acted outside the realm of literary and traditional monastic erudition. The most prominent figure was Polyxena of Lobkowitz (1566–1642), who was an adamant administrator of a vast manor and drew admiration (and hatred) for her negotiating skills. The literature dealing with the history of women in the Czech lands frequently quotes a letter addressed to Polyxena, written by Charles Senior of Zerotin. In the letter, this leading aristocrat refers to Polyxena as the first lady of the kingdom and highlights her intellectual capabilities, which at that time was an exceptionally overt expression of respect (Janáček 1996, 74; Ryantová 2016, 225–227). Polyxena was influenced by her mother, the Spaniard Maria Manrique de Lara, who hosted, together with her other daughters, the so-called salon of Pernstein. Czech aristocratic circles of the time did have gatherings of ladies (*fraucimors*), yet with some exceptions, these gatherings were not based on mixed audiences and sociability. A search of the Czech lands for an institution similar to French salons of the 1600s might uncover only this salon of Pernstein. Its figurehead, Maria Manrique de Lara, was the widow of the Chancellor Vratislaus of Pernstein. Her salon was a centre of Catholic interests and, simultaneously, a gathering place for foreign visitors coming to Prague, especially from Spain and Italy (cf. Kašparová 2010). The hostess and her retinue of ladies were well known for witty conversation and an interest in Catholic politics. Although there is evidence about the functioning of this salon from both local and foreign visitors, no proof of philosophical or scientific interests of the participating ladies has survived, although it can be assumed that the stimulating conversations at least touched upon these topics.

As in the rest of Europe, the Francophonie of the Czech nobility developed in the second half of the 1600s, especially in connection with the political growth of France under the rule of Louis XIII and Louis XIV. While German was used in administrative fields, French became the language used in private. French was spoken in noble families and in salons, and private correspondence was written in French as well. As will be shown in the next chapter, the preserved manuscripts of women authors of that time were predominantly in French, and their translations are most commonly from French as well.

2.3 Baroque and Classicism

Among the most educated women in the Czech territory in the Baroque period were the daughters of Franz Anton von Sporck—elder Maria Eleanora (1687–1717) and younger Anna Katharina (1689–1754). Their father was a major figure of the Czech baroque culture, well known for his interest in literature and art (cf. Preiss 2003). Sporck was interested in developing spirituality and morals; he was keen on studying reformation movements in Christianity, especially in manifestations stemming from Jansenism and Quietism. In line with this moral quest, he decided to translate from

French to German selected works that he intended to spread among his serfs to boost their religious spirit. The publications were intended to simultaneously enhance the status of the Sporck house, which was part of the newly ennobled aristocracy. The books were therefore often dedicated to the members of other houses. Franz Anton commissioned his two daughters to translate the selected works, making use of their immense philological erudition. Simultaneously, he demonstrated the excellent education and skills of his daughters, about whom he cared very much due to the absence of a male heir (cf. Radimská and Lenderová 2002; Čapská 2016, 2017).

The daughters took the first texts with them to translate when they were sent to foreign monasteries in 1701 to be educated: Maria Eleanora to the convent of Annonciades in Rottenbuch near Bolzano; Anna Katharina to a Benedictine abbey in Sonnenburg. To her father's disappointment, Maria Eleanora decided to join the order. To be closer to his daughter, the count established a subsidiary of the monastery on his estate in Chroustníkovo Hradiště and insisted on staying in touch with his daughter despite the strict seclusion prescribed by the Annonciades order. The younger daughter expressed her wish to enter the monastery as well; upon her father's intervention, however, she was forced to marry her cousin Franz Karl Swéerts.

Both sisters translated mainly between 1702 and 1715, when the last work was printed. They were very keen on this activity and even selected some texts to translate themselves. They dealt mainly with religious and philosophical French literature, and they proved to be skilful translators. Maria Eleanora had an active approach to translating; she adapted and supplemented some texts herself, which was common in translating in that period. The books that she translated included Teresa of Ávila's *Contemplation on the Our Father Prayer*, from a French translation. The Preface to the translation is very likely her work; she added her own eighth chapter to the seven texts written by St Teresa. To Pictet's *Traité contre l´indifférence des religions* [*Tract Against Indifference in Faith*], Maria Eleanora added her own chapter, titled 'Which Faith is the Right One', in which she demonstrated her knowledge of St Augustine's works (Čapská 2016, 43).

Maria Eleanora might have chosen herself the translation of a part of *Les consolations de la philosophie et de la théologie* [*On the Consolations of Philosophy and Theology*], a treatise from the French translation of Boethius by René Ceriziers; she dedicated the text to her father. In 1712, Sporck published in Prague a translation of the work entitled *Widerlegung der Atheisten, Deisten und neuen Zweifler* [*Refutation of Atheists, Deists and New Doubters*]. Its author might have been Michel Mauduit, a theology professor in Paris. He was influenced by Blaise Pascal; the first part is more likely a loose adaptation of one of the chapters from Pascal's *Pensées*. According to some scholars, this part was authored by Maria Eleanora herself (Radimská and Lenderová 2002, 139). It is evident that in the supplemented texts, Maria Eleanora freed herself from the passive role of a translator; these activities of hers deserve further research.

In addition to books for her father's printing workshop, Maria Eleonora also translated texts for her order; this work concerned mainly prescriptive texts and instructions of the order. In doing so, she inspired other translators who continued to translate within the order even after her death. The younger Anna Katharina translated especially educational religious texts, e.g. *Warhafte Kennzeichen derer menschlichen Gemüter: Sittliche und Christliche Gedanken* [*Truthful Signs of the Human Minds: Moral and Christian Thoughts*] (1710) and *Christliche Betrachtungen und Sittenlehren* [*Christian Considerations and Moral Sciences*] (1714). The translations of both sisters were abruptly ended by the sudden death of Maria Eleanora and the arranged marriage of Anna Katharina in 1712. However, Anna Katharina and her husband went on publishing books focused on religious topics (cf. Čapská 2016).

The philosophical and scientific interests of Czech women are also evident from the lists of books acquired by private castle libraries. The most distinct example of collecting and reading activity is in the construction of the library in Český Krumlov, which was initiated by Marie Ernestine von Eggenberg, née von Schwarzenberg (1646–1719). Her accomplishments and the Eggenberg castle library were described in depth in the publications of Jitka Radimská (2007). Princess Marie Ernestine had a great command of French—she translated the first fourteen *Letters* by Seneca from French to German; a manuscript of this translation has survived in the library (Radimská 2007, 24). The princess also read French translations from other languages, including from Latin and Spanish.

Marie Ernestine had at her disposal Pascal's *Lettres provinciales*, banned Jansenist books, Bayle's *Dictionnaire historique et critique*, Fontenelle's work *Histoire des oracles*, and even the 'feminist' works by Poullain de La Barre. In the preserved books, mostly marked with the initials of the learned reader, one can find Marie Ernestine's handwritten notes and other evidence proving her thorough reading of the texts and their reception. These often concern general theses related to moralist literature and thoughts that captivated the reader, who intended to return to them repeatedly. For example, in Senault's work *De l'usage des passions*, Marie Ernestine wrote: 'la morte [sic] et aflictions, sont les conditions de la Vie Marie Ernestine Princesse d'Eggenberg 1677' (Radimská 2007, 82). In the library acquisitions, it is also possible to trace Marie Ernestine's growing interest in moralist literature. In a eulogising text from that period, the princess is depicted as Pallas Athena, 'in whom the entire community of scholars has its excellent advocate' and her influence on her husband and brother is also mentioned, albeit within the contemporary rhetoric.[2]

In the course of the 18th and early 19th centuries, the scientific and philosophical interests of other women readers are documented, although they do not equal those of Marie Ernestine. For example, Pauline of Schwarzenberg (1774–1810) was able to read in several languages and was interested in the works of Jean-Jacques Rousseau and in texts from the field of medicine. Rousseau's influence was also evident in her approach to upbringing and childcare (cf. Lenderová 2004).

Further evidence of the interests of female readers are included in the personal diaries that spread throughout the Czech lands beginning in the late 1760s and early

---

[2]'For a woman, Ernestine stands out due to her profound wisdom, as is confirmed by Kristian himself; when he decided about important matters, he used her advice to his own benefit, and so did His Highness Ferdinand b.m., brother of her Highness Ernestine, who will be mentioned further on.' (Radimská 2007, 86, 85)

1770s. The diaries also contain drafts of moral and educational contemplations and reflections on their own status. This interest was related to the new emphasis on the role of the mother, stemming mainly from the ideas of Jean-Jacques Rousseau and his work *Émile* (1762). Ivo Cerman states that the local aristocratic environment in the 1770s had written instructions for mothers from Maria Josepha of Windisch-Grätz and Maria Christina of Dietrichstein; their writings demonstrate an interest in the natural early education of children. At the beginning of the 1800s, Maria Leopoldina of Windisch-Grätz wrote an essay in French entitled *Règles de vie* [*Rules of Life*], intended for her daughter, in which she advocated morals based on Christianity (Cerman 2012, 248, 252).

A cosmopolitan figure connected to the Czech lands was the Russian aristocrat Alexandra Shuvalova (1775–1847), who came from an important noble dynasty. Her father Andrey Petrovich Shuvalov (1743–1789) was a prominent diplomat and a great admirer of Voltaire. In 1765, he even visited Voltaire in Ferney and they stayed in touch afterwards. In 1797, Alexandra married the Moravian count Franz Joseph of Dietrichstein and moved to Vienna. The marriage ended in 1804, and Alexandra sought consolation in faith; she intensively read religious works as well as fiction. The Dietrichstein family archives still hold her correspondence and manuscripts of her works written in French (cf. Musilová 2014). Alexandra was an apt writer, even though most of her texts were unfinished. In contrast to her father, Alexandra disagreed with deism and criticism of Christianity, as is evident from her rejection of the work *Dictionnaire infernal*. As a reader, she preferred moralist literature, yet she also liked novels and poetry. Her philosophical interest is evident especially in her moralising treatise *L'art de s'élever soi-même* [*Art of Self-Education*], which she addressed to her granddaughters, and a fragment from the comedy *Le philosophe moderne* [*The Modern Philosopher*], in which she made fun of the Enlightenment's view of freedom and equality. In her work on education, she addressed general issues pertaining to the meaning of human life and the search for happiness. When dealing with the education of women, she demonstrated a conservative point of view—she thought that women belonged at home. Despite that, she believed that women should have a basic knowledge of various scientific disciplines, in order to be able to hold conversations with others. Although Alexandra Shuvalova's view was affected by her religious conversion, her polemic against Enlightenment views implies she was well acquainted with French works and interested in philosophical topics.

Another author whose manuscripts have survived is Karolina Ferdinandi (1777–1844). Karolina, the daughter of a government clerk from Galicia, was the wife of Franz Adam Waldstein, an eminent natural scientist. Waldstein is said to have nicknamed Karolina 'Aspasia'. She also knew the scholar and adventurer Giacomo Casanova, who worked in Bohemia as Count Joseph Charles Waldstein's librarian at the Duchcov Chateau. He allegedly turned Karolina into one of the heroines of a philosophical dialogue he wrote during his stay in Duchcov (Cerman 2012, 390–391). Karolina and Franz Adam dealt with various literary forms as a leisure-time activity; however, they also revealed more serious philosophical interests in their writings. Franz Adam was keen on Enlightenment literature and translated some texts into German. For example, his adaptation of Voltaire's short story *Zadig* has survived to this day. Karolina herself authored several preserved manuscripts in German and French, including short French contemplations on history and philosophy and an essay on love. Other women from the salons created occasional works of fiction as well (Cerman 2012, 388; cf. Montet 1904).

References

Bránský J (2008) Čtyři z Boskovic. Albert, Boskovice

Čapská V (2016) Mezi texty a textiliemi. (Swéerts-)Šporkové, textové praxe a kulturní výměna na přelomu baroka a osvícenství. Scriptorium, Praha

Čapská V (2017) Maria Eleonora Sporck (1687–1717) and Anna Katharina Swéerts-Sporck (1689–1754): practitioners and promoters of the word at the edge of the Enlightenment. In: Lehner UL (ed) Women, enlightenment and catholicism: a transnational biographical history. Routledge, London, pp 132–148

Cerman I (2012) Šlechtická kultura v 18. století: filozofové, mystici, politici. Nakladatelství Lidové noviny, Praha

Christianson JR (2002) Tycho and Sophie Brahe: gender and science in the late sixteenth century. In: Christianson JR et al (eds) Tycho Brahe and Prague—crossroads of European science. Verlag Harri Deutsch, Frankfurt am Main, pp 30–45

Dinzelbacher P (1993) Mittelalterliche Frauenmystik. Schöningh, Paderborn

Dinzelbacher P (2003) Světice, nebo čarodějky? (trans: Babka P). Vyšehrad, Praha

Hoffmannová J (2016) Prvenství žen. Ženy iniciativní, vzdělané a tvořivé. Ústav T. G. Masaryka, Praha

Janáček J (1996) Ženy české renesance. Brána, Praha

Kašparová J (2010) České země a jejich obyvatelé očima románského světa 16.–17. století. Veduta, České Budějovice

Kavka F (2002) Čtyři ženy Karla IV. Královské sňatky. Paseka, Praha

Klassen JM (1999) Warring maidens, captive wives and Hussite queens. Women and men in war and peace in medieval Bohemia. Columbia University Press, New York

Kopičková B (2002) Žena evropského středověku v zajetí své doby. In: Lenderová M (ed) Eva nejen v ráji. Žena v Čechách od středověku do 19. století. Karolinum, Praha, pp 13–44

Kopičková B (2008) Eliška Přemyslovna. Královna česká. Vyšehrad, Praha

Kosmova kronika česká (2005) (trans: Hrdina K, Bláhová M). Paseka, Praha

Lenderová M (2004) Tragický bál. Život a smrt Pavlíny ze Schwarzenbergu. Paseka, Praha

Muneles O, Vilímková M (1955) Starý židovský hřbitov v Praze. Státní pedagogické nakladatelství, Praha

Muraro L (1987) Vilemína und Mayfreda. Die Geschichte einer feministischen Häresie (trans: Kempter M). Kore Verlag, Freiburg

Musilová M (2014) „Gâtée par le monde." Literární dílo kněžny Alexandry z Dietrichsteina. Master's thesis. South Bohemian University in České Budějovice, České Budějovice

Pelcl FM (1773–1782) Abbildungen böhmischer und mährischer Gelehrten und Künstler, nebst kurzen Nachrichten von ihren Leben und Werken I-IV. Bey Iohann Karl Hraba, der H.H. Stände in Königreiche Böhmen Buchdrucker, Prag. (German version of Latin publication by Voigt, Mikuláš Adaukt. Effigies virorum eruditorum atque artificium Bohemiae et Moraviae, I, II. Prag, 1773–1782.)

Preiss P (2003) František Antonín Špork a barokní kultura v Čechách. Paseka, Praha

Putík A (ed) (2009) Path of life. Rabbi Judah Loew Ben Bezalel. Ca. 1525–1609. Academia, Praha

Montet A, Prévost de la Boutetière de Saint-Mars, baronne du (1904) Souvenir de la baronne du Montet. 1785–1866. Plon, Paris

Radimská J, Lenderová M (2002) Barokní čtenářky: Eleonora a Kateřina Šporkovy, Marie Arnoštka Eggenbergová. In: Lenderová M (ed) Eva nejen v ráji. Žena v Čechách od středověku do 19. století. Karolinum, Praha, pp 131–154

Radimská J (2007) Knihovna šlechtičny. Francouzské knihy Marie Ernestiny z Eggenbergu na zámku v Českém Krumlově. Jihočeská univerzita, České Budějovice

Ryantová M (2016) Polyxena z Lobkovic. Obdivovaná i nenáviděná první dáma království. Vyšehrad, Praha

Sokol J (ed) (1993) Mistr Eckhart a středověká mystika. Zvon, Praha

Stevenson J (2005) Women Latin poets. Oxford University Press, Oxford

Šedinová J (1996) Židovské ženy v Praze v 16.–18. století. In: Pešek J, Ledvinka V (eds) Žena v dějinách Prahy. Scriptorium, Praha, pp 91–100

Špička J (2010) Francesco Petrarca travelling and writing to Prague's court. Verbum Analecta Neolatina XII/1:27–40

Štěříková E (2014) Moravské exulantky v obnovené Jednotě bratrské v 18. století. Obrazy ze života. Kalich, Praha

Storchová L (2003) Alžběta Johanna Westonia – rara avis v humanistické res publica litteraria. In: Westonie AJ (ed) Proměny osudu (trans: Petrů E). Atlantis, Brno

Tiktiner (Bat Meir) R (2009) In: von Rohden F (ed) Meneket Rivkah: a manual of wisdom and piety for Jewish women (trans: Spinner S, Tszorf M). The Jewish Publication Society, Philadelphia

Třeštík D (1966) Kosmas. Svobodné slovo, Prague

Urbánková E, Stejskal K (1975) Pasionál Přemyslovny Kunhuty. Passionale abbatissae Cunegundis. Odeon, Praha

Westonie AJ (2003) Proměny osudu (trans: Petrů E). Atlantis, Brno

Ženy ve vědě do roku 1945 (2013) 3.0 Česko. http://albina.ff.cuni.cz/index.php/Hlavn%C3%AD_strana. Accessed 14 Jan 2019

Chapter 3
1820s–1918: An Expansion of Women's Intellectual Pursuits

Abstract Women in the 19th century Czech fields of philosophy and science were influenced by two main lines of thought. The first was the social-political and ethical ideas of Bernard Bolzano, who rejected the traditional concept according to which women cannot participate in scholarly work, as they are not capable of abstract reasoning; he saw people's social roles as being the ones in which they most contribute to the rationally conceived happiness of the whole. The second intellectual influence is related to Romanticism, which perceives reality not only as a subject of exact research but also as a source of moral and aesthetic emotions. Romantically conceived science claims that the natural differences in men's and women's dispositions can be used to describe natural processes and their driving forces, thus contributing to their deeper understanding. The basic interpretational framework for understanding women's work in philosophy and science in this period is the process by which the modern Czech nation was formed. The concepts and ideals of the National Revival shaped the form and development of the Czech women's movement, the reasons and motives on the grounds of which women entered the public space, the development of women's education, and the answer to the question of whether philosophy and science can bring women personal satisfaction and happiness. Moreover, the contemporary discussion on the legitimacy of using animals for scientific purposes, taking place in the Czech lands independently of the discussion on women's civil and political rights, also had a specific character and development.

3.1 The Needs of the National Revival

The first half of the 1800s is connected with the ongoing process of forming the modern Czech nation, referred to as the Czech National Revival. The core of the process's first phase, which took place into the last third of the 1700s, was local patriotism. Its supporters justified their activities by claiming a deep, emotional relationship with, and love of, the place where they were born or lived. Later, this shifted

Z. Jastrzembská et al., *Czech Women Philosophers and Scientists*,
SpringerBriefs in History of Science and Technology,
https://doi.org/10.1007/978-3-030-61516-1_3

from an emphasis on the love for one's homeland to the love for a nation, the beginning of national patriotism as well as the gradual promotion of the linguistic concept of the nation during the 1820s. The fact that there was a prevailing opinion that science is an integral part of every nation's culture and, thus, the need exists to develop a linguistically Czech science was key for the development and formation of science (cf. Janko and Štrbáňová 1988, 9–25). In the field of literature, Czech revivalists could build on the Czech language of older eras, especially on the humanistic Czech language of the 1500s. However, the situation within scientific discourse was much more complicated. The beginnings and development of many disciplines of modern science occurred during a period when the Czech language was displaced from this plane of communication. To develop a linguistically Czech science primarily meant developing a new vocabulary. The creation of technical scientific terminology started in the 1820s. The founding of new scientific institutions and journals played an important role in this endeavour. The first Czech science journal *Krok* (named after Duke Krok, a legendary figure in Czech history) was founded in 1821. This journal sought to show, by publishing translations as well as original scientific work, that the Czech language is capable of expressing complex and abstract thoughts. Analogous activities were part of the *Časopis českého muzea* [*Journal of the Czech Museum*], which the *Vlastenecké muzeum* [Patriotic Museum] in Bohemia has been publishing since 1827. Scholarly literature written in Czech was also published with the help of the *Matice česká* [Czech Cultural Foundation] organisation, established in 1831.

In the Czech lands, the institutional and organisational foundation for scientific work was relatively narrow. There was the university in Prague (the official name at the time was *Universitas Carolo-Ferdinandea*, or Karl-Ferdinands-Universität [Charles-Ferdinand University]); it had gone through a significant reformation, during which the German language was introduced into the teaching of some subjects during the 1780s. The establishment of German as an educational and official language was part of the reforms made under the Habsburg monarchy by Emperor Joseph II. A united language was one of the tools used for the unification of the Empire. German definitively replaced Latin at the university in Prague during the 1820s. The situation at the university in Olomouc was complicated after abolishing the Society of Jesus (1773), who practically ruled over the fields of education and science at that time. From the 1780s to the end of the 1820s, the university's status fell to that of a mere lyceum (Kostlán and Niklíček 2018, 145). The *Královská česká společnost nauk* [Royal Bohemian Society of Sciences], originally known as *Soukromá společnost nauk* [Private Society of Sciences], continued to operate. The society was established during the 1870s as an expression of local patriotism, through which the nobility wanted, among other things, to promote its own political and economic interests. This local patriotism was manifested in scientific work through a focus on historical research, legal history, cultural monuments, and the language of the Czech lands, as well as the study of nature and natural resources of this region. The society published the journal *Abhandlungen einer Privatgesellschaft in Böhmen zur Aufnahme der Mathematik, vaterländischen Geschichte und der Naturgeschichte* [*Transactions of a Private Society in Bohemia on Mathematics, National History, and Natural History*], although its activities declined at the beginning of the 1800s. A more significant recovery in activity for this society took place during the 1840s when František Palacký (1798–1876), an eminent Czech historian and politician, acted as its secretary. Palacký was successful in reorganising the structure of the society as well as increasing the number of its ordinary and extraordinary members. Members of the Royal Bohemian Society of Sciences, specifically Pavel Josef Šafařík (1795–1861), František Ladislav Čelakovský (1799–1852), Karel Jaromír Erben (1811–1870), and Jan Evangelista Purkyně (1787–1869) became key figures in the revival efforts. The frequency of meetings and lectures increased under the leadership of Palacký. However, he was not successful with his idea that scholarly lectures should be accessible to the public (cf. Pokorná 2018). Palacký focused his research on Czech history. In his extensive work *Dějiny národu českého v Čechách a v Moravě* [*History of the Czech Nation in Bohemia and Moravia*] (1836–1876), Palacký mapped the history of the Czech lands from ancient times to the year 1526, when the Habsburgs assumed the Czech throne. The philosophical concept of Czech history that Palacký presented helped to justify the goals of the National Revival, primarily at its political level (cf. Zouhar 2008, 65–67). The newly emerging provincial museums played an important role in the development of research during the first half of the 1800s. A museum was established in Těšín in 1802, followed by the establishment of the *Slezské muzeum* [Silesian Museum] in Opava in 1814. The *Moravské zemské muzeum* [Moravian Museum] in Brno was founded in 1817, one year after the Patriotic Museum, later named *Národní muzeum* [National Museum], in Prague. The mission of these institutions was not only to cultivate science, but also to spread its results throughout society, popularise these results, and raise public awareness. Therefore, the demand for linguistically Czech science partly emerged from these purely educational bases. The bearer and preserver of the Czech language, at this point in time, was chiefly the rural people, who were lacking in higher education. The problem of understanding and the lack of clarity in establishing scientific terminology was not easy to overcome. The fact that the beginning of the 1800s was also the time of the industrial revolution played an important role in the necessary dissemination of scientific results.

Women, in large numbers, were becoming students and consumers of science in the late 1830s and early 1840s. At around the same time, they were attempting their own creative scientific work in the form of compiling an encyclopaedia. The fact that this compilation was a joint venture appears to be important.[1] Until this time, there are only documents regarding individual cases in which a woman acted primarily as a helper in the scientific work of a family member or person within her close circle of friends. The influence of socio-political events decelerated the development of science during the 1850s. The failure of the revolution in 1848 led to the onset of Bach's neo-absolutism, primarily basing its stability on censorship and police authority. A number of areas of scholarly interest (including natural research) were subject to the watchful supervision of the police. The tense political situation

[1] From the beginning of the 1840s, women also collectively and publicly entered the field of literature, where they published an almanac of their Czech written attempts at poetry *Pomněnky* [*Forget-Me-Nots*] (1843).

loosened at the beginning of the 1860s. Part of the restoration of public associations was the establishment of new trade societies and scientific journals. The *Spolek českých lékařů* [Association of Czech Physicians] was established in 1862 and began publishing the *Časopis lékařů českých* [*Czech Physicians' Journal*]. One year later, the *Komitét pro přírodovědný výzkum Čech* [Committee for the Natural Scientific Exploration of Bohemia] was founded (cf. Sekyrková 1990). The *Jednota českých filologů* [Association of Czech Philologists] and the *Český právnický spolek Všehrd* [Všehrd Association of Czech Lawyers] were established in 1868 and one year later, the *Jednota českých matematiků a fyziků* [Association of Czech Mathematicians and Physicists] was also established and began publishing the *Časopis pro pěstování matematiky a fyziky* [*Journal for the Pursuit of Mathematics and Physics*] (cf. Kostlán and Niklíček 2018, 180–184). The 1860s were also an important milestone in view of the developments regarding the Czech women's movement. Women began entering the public sphere to a greater extent and began founding their own associations and journals. The Association of St Ludmila, established in 1851, was the most significant contributor in the area of charity. The *Ženský výrobní spolek* [Women's Manufacturing Society] (1871), alongside other similar associations that operated in various regions of the country, played the most important role in the development of women's industrial education. The first women's journal, *Lada*, was established in 1861; ten years later, *Ženské listy* [*Women's Journal*] was first published (cf. Bahenská 2005). Before women were able to work in science and philosophy, key discussions took place regarding a woman's tasks in the new social conditions, the purpose an education would serve a woman, what a woman's responsibilities would be, and whether knowledge can bring a woman happiness. During the 1870s and 1880s, women's associations focused on the construction of industrial and business schools that would offer knowledge that was useful in practice. The question of a woman's status in society was seen as a considerable social problem. The activities of middle-class women mainly focused on caring for the poor and ill. For the middle class of this time, the Church continued to have a significant influence on the structure and content of a girl's education. The *Americký klub dam* [American Ladies' Club] became the symbolical centre of intellectual influence. From 1865, the year it was established, until 1889, it organised regular scholarly lectures. During these lectures, prominent Czech philosophers and scientists familiarised women with topics of various disciplines (literary theory, poetry, music, history, physics, astronomy, botany, zoology, medicine, technology, etc.). The American Ladies' Club was also the place where women first publicly presented their own lectures. The Austrian government did not support the construction of secondary schools (grammar schools) for girls and never incorporated them into their public educational system. The first grammar school for girls, Minerva, which was opened in Prague in 1890, is a result of the association's activities. Proper women's university studies were permitted in 1897. The first graduates of universities then joined Czech society at the beginning of the 1900s. This text uses 1918, the year when the first independent Czechoslovak state was established, to frame the work of women in the history of Czech philosophy and science. The establishment of the Republic was important in terms of the evolution of women's rights. The constitution of the Czechoslovak Republic guaranteed equal status for men and women within all realms of life, and the representatives of the women's movement entered into a new era with great expectations and new tasks.

The influence of the ideas of the National Revival can be seen in the work of women in the fields of philosophy and science during the entire outlined period, even though the intensity of this influence weakened over time. The needs of the Czech nation, for which several generations intensely worked, were largely subordinate to the activities of the developing women's movement. In terms of the National Revival, a perspective was created in which the main goal of a woman was to raise nationally conscious children and to be a partner and helper to the man. This perspective influenced debates, even during the second half of the 1800s, when the revival process had reached its climax and the Czech nation had been accepted by the Austrian constitution. The idealised concept of women as governesses and helpers was specially projected into discussions about education, duty, and happiness in the early 1860s, continuing to influence the activities of women during the 1870s and 1880s. Women who entered the public sphere at this time were constantly motivated in their activities, with national interest being the main reason. Apart from the ideas of the National Revival, two other influences of the 1800s acted on the presence of women in philosophy and science: the thoughts of Bernard Bolzano and Romanticism. Bolzano's perspective on the nature of women led to arguments against those who disapproved of women in science on the basis of their supposed inability to use abstract reasoning. Romanticism, on the other hand, provided reasons why women should be actively involved in scientific work. Bolzano's socio-political opinions influenced notions regarding the role of women in society during the second half of the 1800s. On the basis of his ethical principle, according to which it is always necessary to select the option that contributes most to the common good, it was derived that women, given their circumstances, can benefit the Czech nation the most in the field of charity by tending to the poor and socially needy. The character of Czech society, typically quick to reward sacrifice (one's personal sacrifice in the interest of higher objectives or for the common good), and the criticism of individualism started to change during the 1890s. The influence of ideas and concepts, rooted in the National Revival during this time, distinctly lost strength and the Czech women's movement splintered internally. Representatives of the new generation entered the public sphere, benefiting from the success that had been achieved in the field of education and raising demands for civil and political rights. Charlotte Garrigue-Masaryková (1850–1923) contributed in her own distinctive fashion to anchoring thoughts of liberalism within Czech society. Garrigue-Masaryková, together with her Czech language teacher, Marie Blažková, translated and in 1890 published the work *The Subjection of Women* (in Czech *Poddanství žen*), written by John Stuart Mill.

3.2 Women Supporting Men's Virtues and Happiness

One of the most dominant and simultaneously difficult to grasp influences shaping the thinking of a generation of early 19th century Czech society was Bernard Bolzano (1781–1848). Bolzano lectured at the university in Prague for 15 years; during that time he played a role in forming the beliefs of leading figures of the nascent field of Czech science. His students included František Ladislav Čelakovský, Jan Svatopluk Presl, Karel Bořivoj Presl, and Jan Evangelista Purkyně. In 1805, Bolzano was appointed as a chair of religious science (*Religionswissenschaft*), a position that had only been established the year before. Chairs were established at all Austrian universities in order to restore and boost the status of the Catholic Church and faith after the revolutionary events in Europe in the 18th and 19th centuries. In addition to teaching the subject of religion, which was compulsory for all students of the faculty of arts in all three years of studies, this position was associated with the responsibility of delivering educational speeches, known as exhortations (*Erbauungsreden*), which replaced standard sermons. Bolzano's homilies showed him to be a social and religious-reformation thinker and critic. Against the expectations connected to his position as the chair of religious science, Bolzano spoke about the ideas of Enlightenment Rationalism. Bolzano's homilies were very popular. Students took notes, through which Bolzano's ideas spread beyond the university walls.

The topics of Bolzano's homilies varied greatly; however, the most popular were those in which he commented on contemporary trends and problems. One of these issues was the question of the status of women. His homily *Von der Bestimmung und Würde des weiblichen Geschlechtes* ['On the Mission and Dignity of Womanhood'] was written and read on the Feast of the Ascension of Mary in 1810.[2] In this speech, Bolzano rejected the widespread opinion of the time that women were not capable of acquiring deeper knowledge, as they lacked abstract reasoning, and that the only duty and purpose of women's existence was to give men sensual pleasure. Against these voices, which Bolzano characterised as foolish, undignified, and pagan, Bolzano claimed that the female sex had basically the same perceptiveness of wisdom, virtue, and happiness, and that women principally deserved the same rights and entitlements as men (Bolzano [1810] 2007, 134–135). Bolzano viewed the differences in the rights and duties as natural; moreover, he considered this arrangement to be only a temporary state. The different social status of men and women stemmed from the very important natural predispositions of both sexes. The purpose of these predispositions is both the reproduction of the species and mutual furtherance of virtues and happiness. If there were no differences between the sexes, Bolzano argued that it would be impossible to start families, the elementary societal components that make it possible to develop into individuals. On the grounds of the differences in the natural character and predispositions of both sexes (Bolzano notes

---

[2]First printed in 1852 (Dr. Bernard Bolzano's *Erbauungsreden an die Hörer der Philosophie an der Prager Universität, herausgegeben von einigen seiner Freunde*, Prague, 1852, 174–182). The first Czech edition comes from 1884 (Bolzano, B.: *Řeči vzdělávací akademické mládeži, vol. II*, Prague, 1884, 62–67).

men's tendency towards superiority, lack of compliance, and feeling of power and, in contrast, women's patience, moodiness, gentler temperament, and corporal weaknesses), Bolzano explains and justifies the rule according to which man should be the head of family. Bolzano claimed 'we can clearly see that the wife will frequently resist the domination of the husband, but also that the man would in no way allow her rule over him.' (Bolzano [1810] 2007, 137–138) However, this family arrangement did not allow men to make whatever decisions they chose. In a dispute, the man cannot decide solely in his own interest. Bolzano demanded that men act according to what is most beneficial for the happiness of their family as a whole, according to the best of their knowledge. At the end of his homily, Bolzano noted men's duty to show respect to women (especially sons to their mothers). As with all his homilies, 'On the Mission and Dignity of Womanhood' was constructed by Bolzano in order to assert that his opinions on women's nature and social status correspond with both reason and Christian doctrine.

The image of a woman who helps and supports a man in his effort to achieve the happiness of the whole corresponds to the ideal woman in the Czech National Revival. This image reflects the ideas of the Enlightenment and later also Romanticism. In everyday life, this idea could take on various forms: from a woman who simply understands the importance of her husband's activities for the common good of the Czech nation and who takes care of all the everyday chores and children's upbringing, to a woman raising sons and daughters who are patriots, to a woman who is her husband's partner and helper. It is very difficult to establish a more concrete picture of women's activities in any of the meanings mentioned above, as women acted in a closed circle of family or close friends, away from the public eye. In the surviving correspondence and memoirs, however, it is possible to find some evidence proving women's actions as men's partners and helpers in their scholarly work.

First and foremost, there was Anna Hoffmannová (1784–1842), a woman in whose company Bolzano spent 19 years of his life; during that time, he wrote most of his seminal philosophical works. Hoffmannová came from the impoverished aristocratic house of Janotik of Adlerstein. At the age of two, when her father died and her mother remarried, she was adopted by another family. She was raised in a convent. When she was 18, she married the landowner Josef Hoffmann (1776–1851). In 1816, the Hoffmann family moved to Prague so that their sons could study there. Anna met Bernard Bolzano there in 1823, when her 16-year-old daughter Karolina died. At that time, Bolzano had already been dismissed from his university chair due to his open-mindedness. He spent most of the following years—first only in summers, and then after 1830 all year—on the Hoffmann farm in Těchobuz. In this small village in southeastern Bohemia, about 100 km from Prague, Bolzano wrote most of his pivotal philosophical works, including *Athanasia oder Gründe für die Unsterblichkeit der Seele* [*Athanasia or Reasons for the Immortality of the Soul*], *Wissenschaftslehre* [*Theory of Science*], and *Von dem besten Staate* [*On the Best State*].

The main source of information about the relationship between Bolzano and Hoffmannová is the memoirs of her husband, Josef Hoffmann, which he wrote after Bolzano's death and published in 1850. Hoffmann's memoirs show that Anna took care of Bernard with motherly love and looked after him when he was ill. Ever since he was young, Bolzano had suffered from pulmonary tuberculosis, experiencing several life-threatening attacks of coughing blood. In his correspondence with friends, Bolzano calls Anna a 'life saver' and 'benefactor'. However, according to Hoffmann's memoirs, Hoffmannová was also Bolzano's student. Hoffmann recalls Bolzano and Hoffmannová taking frequent walks; he taught her about astronomy and botany, they compiled a herbarium, and he even read to her from his scientific works. Anna's husband states that she was soon able to evaluate and judge the texts herself and that Bolzano thought highly of her judgement and criticism (Hoffmann [1850] 2016, 76–77). In the preserved correspondence, Bolzano writes that Hoffmannová contributed to the birth of his work *Athanasia or Reasons for the Immortality of the Soul*, which was first published anonymously in 1827 and with the author's name in 1838. This treatise contains Bolzano's metaphysical ideas and his teachings about the soul; it is also a consolatory text intended to offer comfort to a grieving person. In what sense did Hoffmannová contribute to this work? Is this particular text based on the long talks that Bolzano and Hoffmannová had after her daughter Karolina died? There is not enough evidence to definitively answer this question; however, certain aspects of her contribution can be inferred if she is considered as the probable audience. Hoffmannová can be assumed to be the intended reader, especially of the chapter entitled 'Renewing a Bond with Our Beloved', proposing evidence independent of God's revelation that after death we will meet our close relatives again. Bolzano tried to prove that it can be deduced on the grounds of reason and experience that after death there will be a certain reunion with those we loved when still alive. Bolzano drew from the idea of soul as a simple substance capable of interaction and improvement of its abilities; in his argumentation, he referred to the laws formulated by contemporary science (e.g. laws of attraction). There is no doubt that Bolzano's text requires an intelligent and educated reader and that Hoffmannová must have been such a reader if the text was intended to provide consolation to her after the death of her daughter. On the basis of the aforesaid testimony of Anna's husband, however, can be inferred that Hoffmannová and Bolzano discussed the immortality of soul,[3] and that her attitudes and opinions could have influenced the final form of *Athanasia*.

The presence of Hoffmannová in Bolzano's life affected his views of women's education and their role in society, as described in his utopian work *On the Best State*. He dedicated his manuscript to Hoffmannová in the early 1830s. Bolzano's vision was that in an ideal society, women would have the same rights as men. In the most efficiently arranged state, no adult citizen should be totally excluded from participating in legislative power. The degree of participation should be primarily derived from one's moral qualities and knowledge of the given issue; no account should be taken of the estate background, property, or sex. Bolzano explicitly stated that women should be elected to the Council of Elders (Bolzano [1932] 2007, 253). This institution played an important role in introducing laws. In an establishment where

[3] Josef Hoffmann explicitly refers to three books that were recommended by Bolzano to Anna Hoffmannová on this topic, which probably became the basis of their discussions (Hoffmann [1850] 2016, 74)

decisions about proposals and rules are made on the basis of universal suffrage, Bolzano claimed that it was right to fear that passions might prevail over reason in the elections. The Council of Elders was obliged to prevent (or enforce) the introduction of laws that contradict (or facilitate) the rationally perceived common good. In his work, Bolzano attributed a crucial role to education and raising public awareness. He regarded their lack as the main source of evil in the world and saw their support as the main factor of progress. In Bolzano's ideal state, there would be compulsory school attendance for both boys and girls up to the age of 14 or 15, including classes in mathematics, geometry, astronomy, physics, chemistry, and natural history (Bolzano [1932] 2007, 279). Subsequent studies at higher level schools would be conditioned only by an individual's abilities and demonstrated love for knowledge. It is interesting to look at the chapter entitled 'On Some Measures Concerning the Differences Between the Sexes', in which Bolzano addressed sex education and the importance of family for society. Bolzano introduced the concept that both men and women should have a vote in citizen decisions. However, if a husband and wife were not unanimous in their voting, they would both lose their vote (Bolzano [1932] 2007, 329). Even though Bolzano's book *On the Best State* was not actually published until 1932, it was well known in certain circles of the Czech intelligentsia nearly 100 years before; as will be shown later, Bolzano's ideas affected the attitudes and activities of some Czech women from 1850 to 1900.

Another prominent female figure, who was later granted membership in a number of scientific societies, emerged in the 1820s in botany. Generally, the research of nature and its sources, which was revived in the Czech lands at the end of the 1700s, was driven by two main motives: theoretical curiosity and economic interests. In the areas where economic interests prevailed, research progressed much more quickly and was also more promoted by the scientific and industrial institutions of that time (typically the area of mineral resources). In contrast, floristic research was in the hands of private researchers for a long time. The aristocracy indulged in flower collecting as a hobby; it was pursued by doctors, pharmacists, keen amateurs, and friends of science. According to the conventions of the time, botany, as 'scientia amabilis', was appropriate also for ladies. Josephine Kablick (1787–1863) was born to the family of David Ettel, the owner of a paper mill in Vrchlabí, a town which is the entrance to the Giant Mountains (in Czech, *Krkonoše*). Josephine had been interested in collecting plants from childhood. At the age of 12, she was sent to Prague to be raised at St Voršila [Ursula] Monastery. She returned in 1806 and married Vojtěch [Adalbert] Kablick (1783–1853), the pharmacist in Vrchlabí, that same year. Vojtěch Kablick was also a competent chemist, interested in ornithology and mineralogy. Together, the couple founded a cabinet of natural science in their house, comprising collections of flowers, insects, fishes, reptiles, birds, and minerals from the Giant Mountains. The theoretical background and systematic form to Josephine's interest in flowers was provided by Václav Blažej Mann (1799–1839), a graduate from the Faculty of Medicine in Prague and a keen botanist. Mann taught Josephine how to preserve flowers appropriately, and he took her to the Institute of Plant Exchange (in German, *Pflanzentausch-Anstalt*). This institute was founded in 1819 by Filip Maxmilián Opiz (1787–1858), a self-taught natural scientist and organiser of floral research in Bohemia. His aim was to arrange the exchange of flowers between collectors. Kablick became a member of this institute in 1825; through her membership, her preparations entered many European herbaria. The contemporary sources greatly appreciated the high quality of the items provided by Kablick via the Institute. Moreover, Opiz nominated Kablick to be accepted to the Royal Bavarian Botanic Society in Regensburg, which she joined in 1841 as a corresponding member. In 1851, Kablick became a member of the natural scientific club Lotos in Prague. Two years later, she was also admitted to the Viennese Zoologic Botanic Society; in 1860, she joined the geologic society ISIS in Dresden. That Kablick was a prominent and respected figure is indicated by the fact that her biography came out in 1849, i.e. when she was still alive. It was written by František Saleský Pluskal (1811–1901), a doctor keen on natural science and history, who dedicated his work on a new method of plant drying to Kablick. Several plants were named after her; the best known is *Petasites kablikianus* (in Czech *devětsil Kablíkové*). Josephine Kablick's name can be found as a separate entry in the *Ottův slovník naučný* [*Otto's Encyclopaedia*], which was the most extensive Czech encyclopaedia of its time.

In the field of botany, two more women who helped and participated in the scholarly work of their family members were the Presl sisters, Karolina Preslová and Terezie Preslová, who were elder sisters of Jan Svatopluk Presl (1791–1849) and Karel Bořivoj Presl (1794–1852). Both brothers devoted themselves to the study of natural sciences and later worked at the university in Prague. Jan Svatopluk deserves much credit for the foundation of Czech professional terminology in many scientific fields. One of his most extensive works is *O přirozenosti rostlin aneb Rostlinář* [*On the Nature of Plants*], which he published beginning in 1820, together with Count Friedrich von Berchtold (in Czech, Bedřich Berchtold of Uherčice; 1781–1876). The work contains engravings with depicted plants. In his recollections of the foundation of the first Czech scientific journal *Krok*, Purkyně writes that the author of these engravings was Jan Svatopluk and they were coloured by Karolina and Terezie (Purkyně 1857, 83–84). At that time, *Rostlinář* represented a unique work in the Czech lands, with a prominent place in the history of scientific illustration.

Czech society long held the idea that women were to support men's virtues and happiness and the desire that men should also support women's virtues and happiness. Even though women's entry into the world of science and culture and their first attempts at creative independent scholarly activity were seen in the late 1830s and early 1840s, women offering invisible support, help, and participation in the work of their husbands continued. For example, there was the wife of the Czech philosopher Ignác Jan Hanuš (1812–1869), Laura Hanušová, née Nádherná (1817–1892), who was born to a Prague family; her father, Jakub Nádherný, owned a tobacco warehouse. In 1838, she married Hanuš, whom she met as her tutor when he was a student earning his living by giving private lessons in the Nádherný family. After they married, they moved to Lviv, where Hanuš worked as a university professor of philosophy for 11 years. Upon their return to the Czech lands, he lectured for a short time at the university in Olomouc, then at the university in Prague, from which he was dismissed in 1852. Hanuš was influenced by the philosophy of German idealism, especially Hegel's philosophy of history, which was one of the main reasons for his dismissal from university. Hanušová helped her husband in his scholarly work the entire time; she read scholarly works and reported about them to him. She composed notes and lists that Hanuš used in his writings (Kučerová 1914, 63).

3.3 Women and Romantic Science

The view of the nature of women that Bolzano expressed in his work can be understood as the qualification that enables the participation of women in scholarly work. Bolzano claimed that women have the same natural perceptiveness for wisdom, virtue, and happiness; therefore, he stated that no rational reason exists for excluding women from the world of science. However, he did not offer any reasons why women should enter science. These reasons are found in the period of Romanticism. The main ideas of Romanticism, which arose as a response to the modern mechanicism and rationalism of the Enlightenment, started to be seen to a greater degree in the early 1800s in nearly all areas of human activity (in literature, music, painting, fashion, etc.) and even influenced the interpretation of science and scientific work in its own specific way. Romanticism is characterised by the universalist union of science and art, which is a prerequisite for the restoration of the unity between humans and nature. For Romantics, nature is not only an object of precise measurement, but also a major source of moral and aesthetic emotions. The role of emotions, fantasy, a sense of mystery, and experiences of surprise during research: these are all factors that naturally opened the doors for women to enter into the world of science and justified their active participation in scholarly work. Proponents of Romanticism came with the notion that individual perspectives are important and that every person can offer a unique perspective to the aspects of nature and, by so doing, can contribute to the understanding of this difficult and complex process (cf. Gascoigne 2003).

The reasons for the participation of women in research, which stands in the background of the Romantic approach to science and nature can be connected to the thoughts of Bolzano, even though Bolzano himself would not likely agree with such a synthesis. The differences in the natural dispositions of men and women can be used and can be beneficial during the process of describing nature and its driving strengths. Romanticism, as the main aesthetic model of the era, influenced the form of the first scientific and philosophical works, which were published in Czech during the 1820s. In their writings, authors used many specific rhetorical and poetic figures, rich metaphors, and other language media with the goal of raising the efficacy of the presented message (cf. Faktorová 2012, 239–261). For the aims of this text, it is primarily interesting to follow the emerging connection between Romanticism and the natural philosophical idea of the unified principle, which governs and pervades all reality. The following paragraphs present an encyclopaedic project that began in the 1840s, the text of Viktorie Paulová, and the travel and natural history work of Božena Němcová. Within these examples, ideas of Romantic natural philosophy stand in the background and play the role of their justification and impulse.

Encyclopaedia for Ladies

The idea to write an encyclopaedia intended especially for women emerged in the *Společnost dívek českých* [Society of Czech Girls], led by Antonie Reissová (1817–1852), also known by her pseudonym Bohuslava Rajská. Using pseudonyms referring to Slavonic roots was a popular practice in the National Revival period, especially among women. Reissová was born as the youngest of three daughters to a German family who moved to southwestern Bohemia in the early 1800s. The family's attitude to Czech culture was affected to a great extent by Reissová's father's friendship with the Czech musical composer Jakub Jan Ryba (1765–1815). In the early 1820s, the Reiss family moved to Prague and hired two young students to teach their daughters Czech. The students were keen patriots who later married Reissová's elder sisters. The eldest sister, Johanna Reissová (1809–1849), married the lawyer Josef František Frič (1804–1876), who co-established the Czech legal terminology. The middle sister, Karolína Reissová (1813–1867), married the doctor Václav Staněk (1804–1871), who made a great contribution to Czech medical terminology. After the death of her parents, Reissová lived with her sisters; in the late 1830s, she started to organise (first at the Staněk home, later at the Frič home) lectures for a group of young women. Contemporary sources show that the lectures were attended by approximately 30 women.

The main lecturer of the women's club was Karel Slavoj Amerling (1807–1884), a physician, natural philosopher, polymath, and pedagogue, who spread his enthusiasm and verve to the people around him and who also carried out risky projects with impressive energy. Amerling's activities were quite varied. He supported publishing original Czech scientific books and translations of world literature into Czech. Within the *Jednota pro povzbuzení průmyslu v Čechách* [Union for Encouragement of Industry in Bohemia], an organisation founded in 1837 in Prague that tried to support the transfer of scientific and technical findings to industrial production, Amerling lectured on Sundays and public holidays for apprentices, journeymen, and craftsmen on various topics, especially those related to chemistry. He also gave educational and mainstream lectures in other towns of Bohemia and Moravia. Amerling was convinced that education was the key factor in the country's progress and wellbeing. In the 1850s, he co-founded the *Fyziokratická společnost* [Physiocratic Society], which, following its French model, strove to acquire more profound knowledge of natural resources and laws and to improve their application in agriculture and industry. Amerling's support of women's education was part of his more general educational and public enlightenment efforts. Amerling saw education as a necessary condition for the development of character and as a way to make it easier for people to cope with their role in society. Hence, regarding women's education, he focused on the disciplines and knowledge that could help women run their households and bring up their children, i.e. knowledge thanks to which women could become better wives and mothers. Above all, Amerling appreciated women's good memory and gift for observation, and he strove by all reasonable means not to leave the other half of humanity 'fallow' but to fully employ women's skills and talents. For example, in 1844, he managed to arrange special lectures for women held at the university in Prague twice a week throughout the entire term. These lectures accordingly pertained to practical chemistry and physics, aesthetics, nutrition, and children's upbringing (Matoušek 1954, 61).

In the late 1830s, Amerling started to build an educational institute in Prague named after the early medieval settlement Budeč. According to legend, Budeč was the centre of education under the Přemyslid dynasty, where Czechs and members of other Slavic nations went to seek education. Amerling conceived his school according to the pedagogical principles of Jan Amos Comenius (1592–1670). Education in Budeč was to be comprehensive, i.e. to include the sciences and arts and crafts, to be instructive, and to be provided in close connection with practice. Amerling's project focused on three main target groups: his school offered education for current and future teachers, craftsmen and workers, and women. Based on contemporary memoirs and plans, Amerling managed to erect a relatively large building featuring not only auditoriums, laboratories, workshops, and a library, but also a printing press, botanical garden, observatory, and hospital offering treatment via purely natural methods. Lectures for women and their activities on preparing the encyclopaedia were moved to Budeč as soon as the school had been built, i.e. in 1843. Women had an auditorium and office of their own at their disposal where they could go any time. They could also use other facilities of Amerling's institute (cf. Jahn 1893, 69–72). The same year the women's club moved to Budeč, Antonie Reissová passed her final teaching exams and opened a private institute for educating girls. The surviving correspondence between Amerling and Reissová indicates that Amerling tried to persuade Reissová to cooperate and move all her activities to Budeč, but Reissová refused to do so. Neither Reissová's institute nor Amerling's Budeč lasted very long. In 1844, after some hesitation, Reissová became engaged to the recently widowed František Ladislav Čelakovský, a prominent poet of the Czech National Revival. A year after their marriage, she left with him for the Prussian Wroclaw, where Čelakovský taught at university. There was no-one to take over her school for girls, which in the meantime had obtained a permit to be run as a public educational institute. The school was reopened in 1845 by Eleonora Jonáková (Kučerová 1914, 33). After Reissová's departure, women's classes were taken over by Svatava Michalovicová, who later became Amerling's wife. Amerling's ambitious project of a comprehensive institute encountered financial problems from the very beginning; it finally went bankrupt in 1848.

The *Encyclopaedia for Ladies* was never completed and was regarded as lost. It was rediscovered in the late 1990s in Amerling's large estate. This discovery has provided a more concrete idea about the work's outline and focus, the form of its individual entries, and its authors. The *Encyclopaedia* was based on the ten-volume German *Damen Conversations Lexikon*, published by Karl Herloßsohn from 1834 to 1838 in Leipzig. Herloßsohn was the pen name of Karel Herloš (1802–1849), a native of Prague, who moved to Germany in the 1820s where he was engaged in literary and journalism activities. In his fiction, which was significantly affected by contemporary Romanticism, he developed themes from important periods of Czech history, such as the Hussite movement. Herloš's historical novels were very popular in the Czech lands, as they helped create the identity of the reviving nation (cf. Urválková 2009).

Most surviving entries of the Czech *Encyclopaedia* were translations or expansions of entries from the German encyclopaedia. The *Encyclopaedia for Ladies* was further extended by Czech and generally Slavonic content and topics from Czech history and topography. It seems that the women from the circle around Reissová drew upon the notes they took during Amerling's lectures when writing some entries. They also had the library in Budeč at their disposal. The correspondence of that time indicates that they were also building their own library. The remaining body of the encyclopaedia is not very large, and only the letters from the first half of the alphabet (A to I) are written more systematically. Today, the manuscript is deposited in the Literary Archive of the Museum of Czech Literature in Prague. Most entries are anonymous; however, some show the author's name. Thus, it is certain that the following female authors participated in the compilation of the encyclopaedia: Reissová, who was very likely its editor, and also Marie Bieblová, Barbara Brádková, Emilie Fryšová, Marie Hošková, Svatava Laadtová, Zdenka Němečková, Josefa Machotková (married name Rudová), Svatava Michalovicová (married name Amerlingová), Františka Svobodová (married name Pichlová), Marie Vidimská, and others. Most of these names have disappeared from the Czech cultural landscape. Basic autobiographical data are available only about those women who were active writers or later published independent works.

When comparing the *Encyclopaedia for Ladies* and *Damen Conversations Lexikon*, the most surprising difference is that the text of the Czech *Encyclopaedia* contains a large number of entries dealing with male figures from the history of philosophy and science. In the *Encyclopaedia for Ladies*, there are individual entries for figures such as Tycho Brahe, Robert Boyle, James Bradley, Michel de Montaigne, Hugo Grotius, Niccolò Machiavelli, Plutarch, Claudius Ptolemy, and Plato. There are no such entries in the German *Lexicon*. In contrast, only a few texts from *Damen Conversations Lexikon*, representing women's figures in the history of philosophy and science, were translated into Czech. Both encyclopaedias include entries on Aspasia, Aedesia of Alexandria, and Maria Gaetana Agnesi. That supports the assumption that the original intention of Amerling, Reissová, and other members of the Society of Czech Girls was to compile an encyclopaedia on their own, and that they started to draw upon translations from the German *Damen Conversations Lexikon* only after the works did not proceed as quickly as originally planned. Even though the text also included two sheets dated in 1852, it is likely that the works on the encyclopaedia did not continue after Reissová's departure for Wroclaw and the end of Budeč.

The philosophical and ideological basis of the *Encyclopaedia for Ladies* project can be approached via the answers to three fundamental questions: Why did they choose the encyclopaedic form? Why should the encyclopaedia be particularly for women? Why should it be written by women themselves? The answers to these questions are interconnected and complement each other. One of the main aims of encyclopaedic works is to summarise and categorise the knowledge that is available at the given moment. The purposes and methods of organising knowledge in encyclopaedias have changed over time. While for example the early medieval encyclopaedias were conceived as paths to education, early modern encyclopaedias primarily show an effort to exclude knowledge lacking an empirical basis. The *Encyclopaedia for*

Ladies developed from the ideas forming the background of 18th century encyclopaedias, i.e. it was not intended to show how various topics and disciplines are interlinked; it presented the individual entries in alphabetical order; it included the fields of science and arts; it paid attention to findings applicable in practice (cf. Yeo 2003). The effort to compile a general encyclopaedic dictionary written in Czech can be traced back at least to the 1820s; it was not completed until the second half of the 1800s (the *Riegrův slovník naučný* [*Rieger's Encyclopaedia*] was published between 1860 and 1874). The first published journals were also of an encyclopaedic character (*Krok*, *Časopis českého muzea*). The desire of Czech researchers to have their own encyclopaedia was related to the introduction of a new Czech terminology and reflected the need to spread general education and special scientific findings among craftsmen and industrialists who could apply them in practice. In Amerling's Budeč, work progressed on as many as three encyclopaedias. Each of them had its own target group, the needs of which were taken into account in selecting the entries. There was a great encyclopaedia aimed at intelligent and educated classes; a small encyclopaedia that was to concentrate on knowledge usable in manufacturing, focused particularly on craftsmen and industrialists; and finally, the *Encyclopaedia for Ladies*. The need to educate women went hand in hand with the specific role that was ascribed to them in the National Revival process, in which Czech women were to bring up patriotic children. When describing the work on the *Encyclopaedia*, Vlasta Kučerová (born 1890), one of the first authors to deal with the history of women's movement in the Czech lands, emphasises the educational character of the entire process. She mentions that women wrote papers on assigned topics and subsequently read them at joint meetings. Thus, they increased their knowledge of the given topics and practised procedures and skills for working on the encyclopaedia. Amerling and his friends then corrected and revised their papers (Kučerová 1914, 35). Another possible explanation of why women were involved in the compilation of the *Encyclopaedia* is related to the contemporary Romanticism. The influence of Romanticism is evident in the surviving body of the *Encyclopaedia for Ladies* not only due to the use of applied means of expression and the way the individual entries are compiled, but even in their selection. For example, there are many historical and mythological themes as well as descriptions of exotic places.

It is interesting to compare the entries on astronomy, which are much longer in the Czech *Encyclopaedia* and reflect some of the basic principles of the Romantic perception of science. The first part of the text, which deals with the history of the field, is a faithful translation of the German original. This part is written in a relatively matter-of-fact language and acquaints its readers with the major figures of astronomy from Ptolemy, Galileo, and Kepler to Herschel and Laplace. While the German original ends with a mention of Joseph Frauenhofen's 'eternal merits', the entry in the Czech *Encyclopaedia*, written by Barbara Brádková (1819–1907), continues further. The author depicts the connection between astronomy and other scientific disciplines (physics, optics, mathematics), and draws attention to the need to revise the knowledge gained through sensory observation. Brádková describes the universe as both an object of precise study and one of aesthetic experience. In the dictionary entry, the beauty of the universe is described by means of various poetic devices and figures (metaphors, personification, comparison, etc.). After listing the planets and other bodies of the solar system that can be seen with binoculars, the author focuses on the topic of the position of human beings in the universe. She emphasises people's privileged position between the macrocosm and the microcosm, which allows us to perform astronomical observations and reveal the mystery of the cosmos. Brádková describes human beings as 'God's favourite', resembling God in their ability to create (people create systems of science and art, transform nature, and build cities and towns).

Natural Scientific Interest

The ideas of Romantic natural philosophy also presented an important stimulus for the research work of Viktorie Paulová and Božena Němcová. Both of these women belonged to different generations and came from different environments. Their lives fatefully crossed within the context of the repressive political system of Bach's neo-absolutism.[4] This subchapter focuses on another shared connection. In the early 1850s, Jan Evangelista Purkyně entered the lives of Paulová and Němcová and decisively influenced the form and direction of their technical interests. Purkyně is regarded as being one of the greatest Czech scientists of the 1800s. He captivated the wider scientific community with his dissertation *Beiträge zur Kenntnis des Sehens in subjektiver Hinsicht* [*Contributions to the Knowledge of Vision from a Subjective Point of View*] (1819), in which he defined his project of experimental research on the physiological foundations of sensory phenomena.

Purkyně was convinced that even subjective sensory phenomena, such as those for which the external world has no answers, have a certain objective quality that can be investigated by scientific means. Purkyně also simultaneously thought that this research task was solvable by strict abstraction and experimentation on himself. He studied visual and auditory phenomena, the physiological groundwork of equilibrium, the effects of medicines on mental functions, the symptoms of intoxication, physiology of speech, and much more. After finishing his studies, Purkyně worked as an assistant dissector for four years at the Faculty of Medicine at the university in Prague. He left in 1823 for the University of Wroclaw, where he was appointed a professor of physiology and pathology. In the early 1830s, Purkyně shifted his professional focus to examining the structure of plant and animal tissues, made possible thanks to the development of microscopic technology. His results contributed to the formulation of cell theory (cf. Kruta 1971). After eight years of hard work, Purkyně successfully established an experimental physiological institute in 1839, which was one of the first workplaces of its kind in Europe. While working in Wroclaw, Purkyně never lost touch with events taking place at home. He later returned to Prague, where he re-joined the university in 1850 and became intensely involved in patriotic activities. He focused a large part of his efforts on the research of natural science, which was not yet systematically supported in the Czech lands. Together with geologist Jan Krejčí (1825–1887), Purkyně established the natural science journal, *Živa* [after

[4]Research into secret police documents revealed that Paulová was a police informant. Her main objective was to report on Němcová's activities.

the Slavic goddess of life and fertility], in 1853. Its main objectives were to offer summarising and informative articles that would be comprehensible to the general public. For the further development of Czech science, his reflection on national academia was of great importance. Purkyně published his ideas between 1861 and 1863 in *Živa*, on the basis of which the *Česká akademie císaře Františka Josefa pro vědy, slovesnost a umění* [Emperor Franz Joseph Czech Academy of Sciences, Letters and Arts; later named the Czech Academy of Sciences and Arts] was founded.

Purkyně's view of nature was influenced by the thoughts of German philosophers, with whom he became familiar during his studies. Purkyně read the works of Kant, Fichte, Schelling, Novalis, Hegel, and Goethe (especially Goethe's book *Zur Farbenlehre* [*Theory of Colours*]). Purkyně primarily expressed his opinions on natural philosophy in his studies published during the 1850s in *Živa*; however, there are even mentions as early as in his dissertation, in which he defended the method of analogy as a favoured and quintessential method of examination for those studying natural philosophy. In his doctorate, Purkyně pointed out the analogy between individual sensory phenomena and the analogy between sensory phenomena and phenomena of the outside world. The most distinct analogies are his opinions on natural philosophy, evident in his work *Papierstreifen aus dem Portefeuille eines verstorbenen Naturforschers* [*Fragments from the Notebook of a Deceased Naturalist*] (1850), which is a typical example of this period's Romanticism in its form and language. It is here that Purkyně outlines the history of science and its individual disciplines as a process of becoming familiar with the Earth as an organism that can feel and think. He believed that before this time, the individuality of the Earth was primarily examined from a spatial perspective (geography, botany, etc.), and that it was much more difficult and important to create a picture of the Earth's time individuality, realising how the Earth has evolved over time and how higher forms of organic life have gradually appeared (geology). In addition to this Earthly body, its psychic or mental manifestations, the main bearer of which are humans, are also the subject of examination. Ethnography investigates the mental manifestations of Earth from a spatial perspective, and history looks at it from the perspective of time. Based on the analogy from the experimental examination of subjective sensory phenomena, Purkyně speculated about how the Earth's subjectivity could be examined by sensory organs (for example, soil as the sense of touch, water as the sense of taste).

Purkyně pointed out the pitfalls of excessive specialisation in science, which prevents the creation of a synthesised and united picture of the world. Purkyně was convinced that the people's perspective could be very useful for interpreting and understanding natural events; he specifically meant the sort of people who were not burdened by any special education, as well as the perspective of women, who have a more developed touch for perceiving the emotional and aesthetic aspects of reality. Purkyně actively supported women in their interests regarding scientific questions. He loaned them books and corrected them in their first attempts at writing expert scientific texts. Purkyně's correspondence indicates that Viktorie Paulová (1834–1856) wanted to send him her texts on natural science for his evaluation in the mid-1850s. Paulová wanted to use these texts and publish a book for young people, *O životě nerostů* [*On the Lives of Minerals*] (cf. Janko and Štrbáňová 1988, 245).

Paulová was born in Prague. Her father owned a factory that manufactured chemical agents. Her family was able to provide her with a good education thanks to their favourable financial situation. Paulová attended the lectures of Karel Amerling and the private institution of Eleonora Jonáková. She was forced to work at the age of 15. She made a living as a private teacher of the Czech language in the family of a police council member in Prague—Leopold Saher-Masoch von Kronenthal. She later opened her own private educational institute for girls. Her requests for the establishment of a public Czech school for girls in Prague were not heard due to her young age. In a brief biography written by her sister Nina, which was published after Paulová's death, Paulová is depicted as a very talented young woman who was primarily interested in botany and natural science (cf. Paulová 1862). Her life changed dramatically after her father was arrested during the events of the revolution in Prague in 1848. When her mother died shortly after, Paulová, as the oldest of her siblings, was forced to take on the responsibilities of her entire family. She died of tuberculosis at the age of 22.

In 1861, the women's journal, *Lada*, printed her text entitled *Věk člověka* ['The Age of the Human']. Here, Paulová described a human's bodily and mental characteristics in individual evolutionary phases. From today's perspective, these writings could be placed in the field of developmental psychology. Paulová's text is framed in natural philosophical ideas about the unity of the universe, in which the spirit of nature pantheistically dominates. Every natural event, including the life of a human, is tied together by essential laws and is aimed at fulfilling its purpose. In this context, Paulová uses the term 'vyvinování' ('transformation'), which natural philosophers used to express the idea of evolution before Charles Darwin. According to Paulová, humans are exceptional in that they are the only creation that have been able to control nature and use nature for their own benefit. Humans have been gifted with a higher power and are thus called to freedom. In her text, Paulová devotes the majority of her attention to the description of psychological characteristics and their differences in men and women in individual evolutionary phases. She also takes note of their influence on choosing a career path. Paulová's text contains many interesting observations and comments. For example, in the context of contemporary debates on the theory of mind, there is a passage in which Paulová devotes her writings to the social development of children and asserts that it is important for a child's correct development that the child is capable of understanding other people as conscious beings.

The ideas of Romantic natural philosophy also served as the initial impulse for the birth of the travel and natural history texts of Božena Němcová (1820–1862), the famous Czech author. Němcová was born as Barbora Novotná, in Vienna, but she was raised in the countryside in Ratibořice, a small village in north-eastern Bohemia, from a very young age. Her parents chose her husband, Josef Němec (1805–1879), for her. Němec worked with the financial guard. Their wedding took place in 1837. Due to her husband's work, their family was forced to move often. Němcová first came to Prague in 1842, where she established contacts with a Czech patriotic society. Here, she befriended Reissová and several physicians and natural scientists who became her teachers (Janko and Štrbáňová 1988, 119). Following a German model, Němcová translated several entries in botany for the *Encyclopaedia for Ladies*. During her stay in Domažlice in southern Bohemia, Němcová wrote short ethnographic texts, e.g. *Obrazy z okolí domažlického* [*Pictures from the Surroundings of Domažlice*] and *Selská svatba v okolí domažlickém* [*A Rural Wedding near Domažlice*], in which she depicted the folk traditions and customs.

Czech revivalists derived their interest in folklore from the belief that the spirit of the nation appears in cultural manifestations and that the pure Czech essence should be sought in the Czech countryside. During the 1850s, Němcová took several trips to Slovakia and Hungary. Purkyně, who financially supported Němcová and also loaned her books, encouraged her to pay greater attention to nature during her travels. Němcová arranged her findings and experiences from her travels in travel and natural-historical texts (e.g. *Vzpomínky z cest po Uhřích* [*Memories of Travels Through Hungary*], *Uherské město Ďarmoty* [*The Hungarian town of Ďarmoty*], and *Kraje a lesy na Zvolensku* [*Regions and Forests in Zvolen*]).

The genre of the travelogue became part of the scientific discussion in the second half of the 1700s, when what were originally discovery voyages became research-focused expeditions. The expeditions of James Cook (1728–1779) were typical of this period. Astronomers, cartographers, geologists, botanists, and zoologists took part in these expeditions; with the help of various instruments (e.g. sextants, chronometers, and microscopes), they were able to perform many varied observations, measurements, and experiments. The messages from their travels offered thorough descriptions of the nature of the lands they visited as well as the local customs and attractions. These texts became more attractive for a wider circle of readers (cf. Iliffe 2003, 618–645). The new generation of explorers, whose most famous representative was Alexander von Humboldt (1769–1859), focused on the exact mapping and surveying of the world, but also on capturing experiences and mediating the emotions the explorer experienced during his travels. Humboldt's works were very popular within the Czech milieu. Reports and translations of examples from his works, especially from *Ansichten der Natur* [*Views of Nature*] and *Kosmos – Entwurf einer physischen Weltbeschreibung* [*Cosmos: A Sketch of a Physical Description of the Universe*], were published in Czech journals beginning in the 1820s. In 1828, Humboldt was elected an honorary member of the Patriotic Museum and five years later as an honorary member of the Royal Bohemian Society of Sciences. In 1848, the university in Prague awarded him two honorary doctorates on the occasion of its 500th anniversary (cf. Haubelt 1996, 38–53).

Humboldt's approach inspired several Czech explorers and was reflected in their books of travels (Karel Slavoj Amerling, Milota Zdirad Polák). In Němcová's travelogues and natural-historical texts, however, the romantic and emotional approach gradually vanished. In her largest text on natural history, *Kraje a lesy ve Zvolensku* ['Regions and forests in Zvolen'], published in 1859 in the *Živa* journal, Němcová explores the region of central Slovakia, describing its countryside in detail, with all its features and components. She enumerates mountains, rivers, the composition of forests, and fauna and flora species. In some cases, she includes their Latin names alongside the names used by the local people. These enumerations are sometimes accompanied by scientific explanations of natural processes. Němcová also discusses the demographic composition of the population, its religion and social structure, economic activities, trades and crafts, and local traditions and customs. Her perspective is very realistic and almost free of emotions. Although she conveys to the reader the electrifying experience from a bath in a natural spring, her text lacks the traditional romantic reflections on the feelings experienced when in direct contact with nature. This emotional distance is probably related to the fact that Němcová also pays attention to poor social conditions that defy the romantic method of presentation. In the descriptions of the locals' traditions and customs, Němcová also mentions their poverty, the lack of education, and the hard work in mines and forests that was often done by children. Social issues were an important topic for Němcová, and they were often included in her literary works.

3.4 Education, Duty, and Happiness

The political loosening that followed the fall of Bach's neo-absolutism led to a flourishing of societies and associations in all areas of life. New scientific societies, clubs, and specialised journals started to emerge in the 1860s, including exclusively women's associations and magazines. The first Czech women's magazine, *Lada*, named after the Slavic goddess of love, marriage, and spring, was launched in 1861. The first issues included texts on scientific topics, such as information about the comet observed that year and about the expedition to the North Pole that was being prepared by the Swedish Royal Society of Science. Nevertheless, scientific news soon disappeared, and the magazine published mainly literary texts, poetry, and fashion news; it also presented important female figures from Czech and world history and scientific information (mainly from the field of chemistry) that could be useful in maintaining a household.

The idea that being a good wife and taking care of a household are the main duty of every woman was influential in the Czech society throughout the 1800s. Public activities were tolerated in young unmarried women; married women could only participate after having accomplished their duties at home. This idea was explicitly expressed in the works of Magdalena Dobromila Rettigová (1785–1845), née Artmannová. Rettigová was of humble origins. Her father and all of her siblings died when she was little. She lived alone with her mother and had to work to sustain the family. In 1808, she married Jan Alois Sudiprav Rettig (1774–1844), a lawyer and Czech patriot, who brought her to writing. Rettigová is mainly known as the author of cookbooks and short stories reflecting moral principles for young girls. Her most well-known cookbook, *Domácí kuchařka, anebo pojednání o masitých a postních pokrmech pro dcerky české a moravské* [*Home Cookbook, or a Treatise on Meat and Fasting Dishes for Czech and Moravian Daughters*] (1826), was very popular and published in several reeditions. Rettigová's opinion on how to be an ideal wife is presented in the book *Mladá hospodyňka v domácnosti, jak sobě počínati má, aby své i manželovy spokojenosti došla* [*What a Young Housewife Should Do to Satisfy Herself and Her Husband*] (1840). Her view was inspired by the virtues attributed to women by Catholicism. A woman was supposed to be hard-working, orderly, and committed to God and her husband. Rettigová gives girls advice and tips on housekeeping, cooking tasty meals, and using natural resources to treat illnesses. Rettigová believed that following this advice would lead to an orderly home and a happy husband who abstained from drinking, gambling, and prostitutes, because all his needs and expectations were met by his family. Rettigová believed that hard work and the willingness to submit were the main ways that men could show respect for women (cf. Neudorflová 1999, 12–19).

A similar view of a woman and wife that justifies the differing roles of women and men in the family by referring to the Christian notion of female virtues is present in the works of Honorata Zapová, born Wiśniewski (1825–1856). Zapová came from a Polish noble family, her father was a nobleman in Galicia (today's Poland and Ukraine). At the age of 16, she married Karel Vladislav Zap (1812–1871), a clerk and later high school teacher, who was interested in geography, history, and archaeology. After the family moved from Lviv to Prague in 1845, Zapová became engaged in the patriotic movement. She participated in the creation of the *Spolek Slovanek* [Association of Slavic Women] in 1848, based on the popular contemporary idea of cooperation and unity of Slavic nations. One of the main aims of the association, whose members were mainly women organised around the figure of Bohuslava Rajská at the beginning of the 1840s, was to open a school for girls. Zapová was very good at gathering money to fund the school. In cooperation with Josef Wenzig (1807–1876), a teacher and supporter of higher education for girls, they even drafted the curriculum. However, in 1850, the Association of Slavic Women ceased to exist (cf. Kučerová 1914, 76–81). Zapová was active as a translator (she translated theatre plays from Polish) and wrote short ethnographic texts acquainting Czech readers with the traditions and customs of her native country. However, her desire to found a girls' school did not fade. In 1855 she finally succeeded, but had to shut the school down after several months due to her illness. She died of tuberculosis at the beginning of the following year. Zapová outlined her opinion on the role of women in the family and their duties, virtues, and education in the book titled *Nezabudky: dar našim pannám* [*Forget-Me-Nots: A Gift to Our Maidens*] (1859), based on the Polish book *Pamiątka po dobrej matce, czyli ostatnie jej rady dla córki* [*Remembrance of a Good Mother, or Her Final Advice to Her Daughter*] (1819) by Klementyna Hoffmanowa (1798–1845). Zapová mentions religion, hard work, patience, gentleness, obedience, gratitude, and modesty as a woman's principal virtues. Girls should receive an education that will help them manage a household appropriately. They should mainly be able to cook and sew and know basic arithmetic. Quite interestingly, Zapová also expected these skills from women from higher social classes, as she believed that they needed to understand housekeeping to prevent their servants from cheating them. She considered cooking and sewing classes to be more important and useful than learning to sing, dance, draw, or speak foreign languages. She also presented a very pragmatic attitude to marriage. She recommended that her readers should marry for convenience and friendship, rather than for love, which for her was associated with suffering and fades away at the end anyway. She saw the failure to marry as something that contradicted the laws of nature. In her text, she also enumerates negative female qualities (backtalk, gossip, vanity, flirting, jealousy, and laziness) that discourage potential husbands. The number of reeditions (the fourth one was released in 1870) suggests that Zapová's book was fairly popular, although it is not clear to what extent Czech mothers really relied on its advice when bringing up their daughters. Although the content of girls' education evolved over time, the argumentation referring to religious beliefs and Christian virtues was recurrent in the ongoing debate, as there was a concern that educating girls might diminish their belief in God. In his letters to Reissová, Čelakovský admits that he could not love a foreigner or 'atheist' and mentions that he was happy to find out about her relationship with religion. In her reply, Reissová reassured him that she was convinced that life should be based on religious faith, although she had learned to think about religious matters in a rational way (cf. Rajská 1872, 156–159).

Knowledge and Happiness

The view of female education presented by Zapová in *Nezabudky* is derived from the belief that unlike men, women cannot find happiness in knowledge. Happiness is seen by Zapová as the main goal of life and, as she explains in the last chapter of her book, she wrote down her experience and advice to show girls how to achieve happiness, which she defines as feeling satisfaction with the status quo, accepting one's fate, and realising that things cannot be changed. This peace of mind or 'paradise in the heart' is understood as an accomplishment of the goal, fulfilling the purpose for which humans were created. In her view, restlessness and unease are associated with idleness; therefore, she always emphasises activity and hard work, which she considers to be a source of happiness. Unmarried girls and women from higher classes who are less burdened by housework and farming are advised to use their free time for reading. The readings recommended by Zapová include books that enhance faith and virtue, theatre plays that teach morals, and travelogues. She also thinks that women should know the history of their nation and its traditions. Philosophical and scientific works are not recommended for girls; Zapová also warns against entertaining books and eccentric foreign novels that can be dangerous for a woman's mind, as they emanate from different habits and cultural traditions (cf. Zapová 1863, 66–70).

The belief that a woman without a higher education has a better chance of being happy and satisfied than an educated one started to change in Czech society in the early 1860s. This idea was explicitly expressed in the works of Josefína Rudová, née Machotková (1806–1877). Rudová belonged to a circle of women that formed around Reissová in the 1840s. She attended the lectures given by Amerling and contributed to the *Encyclopaedia for Ladies*. Reflections on happiness and how to achieve it are the opening to her book *Loterie: hrst naděje a pytel nejistoty* [*Lottery: A Handful of Hope and a Bag of Uncertainty*] (1863) exploring the history and nature of betting. Rudová's understanding of happiness is similar to that of Zapová. To her, happiness is a feeling of satisfaction, relative in terms of external circumstances and conditions. Throughout her text, Rudová uses quotes from Comenius' work *Labyrint světa a ráj srdce* [*Labyrinth of the World and Paradise of the Heart*] (1631), a story of a pilgrim who explores the world seeking happiness only to finally find it inside his soul. Referring to Isaac Newton, Rudová believes that a safe way to achieve happiness is to intensively dedicate time to a certain topic, exploring things and thinking about them. The only thing one needs is a strong will, patience, and courage (Rudová [1863] 1876, 11). In her book, Rudová demonstrates a good knowledge of probability calculations, which was quite unusual at that time. Besides her explanation of the history of betting, she also provides detailed mathematic calculations to prove that the betting person bears all the risk, while the owner of the lottery can be sure to earn a profit.

Rudová's book also includes a critical description of the social consequences of betting. She notes that betting makes poor people even poorer, while the rich have it as a form of harmless entertainment. She reflects on how the world would look if people paid taxes with the same eagerness as they gamble. As the certainty of profit lies on the side of the lottery owner, she believes that lotteries should be operated by the state, who should use the profit for charitable purposes. The reasons that led Rudová to write her book draw on the enlightened idea that education can drive the progress of humankind. Rudová was convinced that her rational argumentation and the analyses provided would make people stop betting, as they would then understand that the chance of winning is small and that betting cannot make them richer or happier. She considered it important to pay attention to the social consequences even when writing about topics that belonged to the field of natural sciences at that time. In her book *Tabák kuřlavý a šňupavý* [*Tobacco for Smoking and Snuffing*] (1864), Rudová explores tobacco planting, as well as its history, chemical composition, and use in pharmacology. In her description, she enumerates different kinds of tobacco, but she also refers to applicable laws, data on the volume of the tobacco industry, and the impact of tobacco on the country's economy and social life. When investigating this topic, Rudová must have studied a large number of resources (her description is accompanied by Seneca's and Cicero's opinions on human health), to which she had access thanks to her husband, Josef Ruda (1807–1875). Ruda studied at the Faculty of Medicine of the university in Prague and was also interested in archaeology, botany, and phytopathology. From 1842, he worked at the Patriotic Museum, first as an assistant and later as a custodian and librarian.

Social criticism was also present in the works of Czech woman writers. Critical descriptions of social conditions can be found in stories and novels by Božena Němcová, Karolina Světlá, Eliška Krásnohorská, and others. Czech readers became acquainted with the ideas of French utopian socialism and communism mainly thanks to the work of František Matouš Klácel (1808–1882), a philosopher and religious thinker. In the revolutionary year of 1848, he published the book titled *Listy přítele k přítelkyni o původu socialismu a komunismu* [*Letters Written by a Friend to His Female-Friend on the Origin of Socialism and Communism*], addressed to Němcová and originally published in a Brno daily. The unfavourable economic situation in the Czech countries influenced the nature and development of the feminist movement. Throughout a significant part of the second half of the 1800s, the so-called 'women's issue' was basically seen as a social one. Most women's associations were founded in order to help the poor and needy, and their activities in the field of education focused on the creation of industrial and business schools (cf. Bahenská 2005). Practical courses were offered, for example, by *Ženský výrobní spolek* [Women's

Manufacturing Society] established in Prague in 1871, and by the *Vesna* association, originally formed as a women's choir in Brno in 1870. Sewing, bookkeeping, and porcelain painting courses were intended for women from lower social classes. Women from higher classes mainly linked their role in society to charitable work.

Marie Riegrová (1833–1891), daughter of František Palacký, was one of the leaders of charitable activities in Prague. Her approach substantially changed the nature, motives, and objectives of the care that was necessary in the second half of the 1800s. Riegrová was inspired by the ethical and social-reformist opinions of Bernard Bolzano, introduced to her by František Schneider (1794–1858), one of Bolzano's pupils. Schneider taught religion at the Technical University in Prague and also worked as a teacher in Palacký's family. Palacký established a closer relationship with Bolzano in the early 1840s, when Bolzano settled in Prague after the death of Anna Hoffmannová. In 1839, Palacký was elected secretary of *Královská česká společnost nauk*, or KČSN [Royal Bohemian Society of Sciences], and introduced changes in order to revitalise the activity of this scientific institution. Bolzano, himself a member of KČSN since 1815, supported these changes and actively contributed to their implementation. Bolzano presented his opinions on charity in a series of exhortations delivered at the university in Prague in 1811. His speeches were first published anonymously in 1847. When considering the motivation for charity, Bolzano emphasised that apart from compassion, the decision to do charitable work must also involve the idea of duty, an awareness of doing what should be done (Bolzano [1847] 1951, 47). Bolzano believed that to deserve being referred to as good, all actions must comply with the highest laws of ethics. As for charitable activities, this means that people should investigate who needs the most help and what means should be employed to achieve the maximum effect. Bolzano thought it important to join forces (by establishing public charitable organisations) and create opportunities that would help the poor to earn their own living (cf. Bolzano [1847] 1951, 37–46). Riegrová put these principles into practice through her own charitable activities. In 1864, she joined *Spolek sv. Ludmily* [St. Ludmila's Association], which had been founded in 1851 and was the only official women's association in Prague until the mid-1860s. *Spolek sv. Ludmily* originally focused on material support for widows and poor girls who could attend its sewing courses. Riegrová introduced several substantial changes. First, she started to assess the necessity and impact of the aid provided. Regular visits to poor households were organised to advise families on sanitary measures and children's education in order to ensure their 'moral development'. Alms were replaced with poverty prevention. An industrial school for girls was established, offering courses in tailoring, accounting, bookbinding, and graphic design, as well as porcelain and glass painting. The products manufactured in the classes were sold and the students were paid. After they finished their studies, the school helped them find a job (cf. Bahenská et al. 2011, 182–190; Červinková 1892, 45–58). Riegrová's work was continued by her daughter Marie Červinková (1854–1895), who carried on with practical activities and projects and also developed theoretical reflections on charity. In her life, Riegrová only published smaller texts in journals, describing the different forms of charitable work and the organisation of industrial schools for girls in France, and comparing the French and German systems of care for the children of poor and working women (kindergartens). A systematic overview titled *Ochrana chudé a opuštěné mládeže* [*Protection of Poor and Abandoned Youth*] was written and published by her daughter in 1887. Červinková parted from the hypothesis that the main criterion for assessing a society's progress and maturity is its approach to women and children rather than the degree of development of science and education. Following a brief historical overview and detailed mapping of the situation in the area of care for people in need in different European countries (France, England, Switzerland, Germany, Belgium, Holland, Austria, and Russia), she surveyed the basic objectives and tasks faced by charitable activities at that time and formulated key principles that could develop and improve charitable work. Červinková is also the author of Bolzano's first biography written in Czech, which she wrote on the occasion of his 100th anniversary (1881). She used Bolzano's autobiography and the memories already published by his friends as the main sources. She also had access to Bolzano's biography by Řehoř Zeithammer (1800–1881), who wrote it for KČSN after Bolzano's death, but left it unpublished because he considered it unfinished. Throughout a large part of her short life, Červinková kept detailed diaries that provide an idea of how the text on Bolzano was being created and what she thought important to include (cf. Červinková 2009, 124–178). Her diaries show that one of her main objectives was to complete the picture of the process with Bolzano and describe the role of each of its participants based on documents acquired from the national archives in Vienna. She also wanted to show Bolzano's relationship to the Church and religion and explain his notion of faith that is 'governed by the principles of reason'. Červinková repeatedly discussed these questions as well as other aspects of Bolzano's work with his pupils and contemporaries. Bolzano's biography by Marie Červinková is more than just a collection of facts. It also contains descriptions and characteristics of Bolzano's work, including those that had not been published at the time.

The *Americký klub dam* [American Ladies' Club], one of the most influential women's associations in Prague, organised charitable activities and lectures on different topics from the fields of philosophy, science, and the arts. The creation of the club in 1865 was prompted by Vojta Náprstek (1826–1894), a patron and promotor of women's emancipation, who returned to Prague in 1858 from America, where he had escaped to avoid imprisonment for his participation in the revolutionary events of 1848. Náprstek was a keen admirer and promoter of technical novelties. He organised exhibitions presenting machines for craftsmen's workshops as well as household equipment and appliances (such as sewing machines and refrigerators) that were invented to make housework easier for women. In his lectures, he talked about the need for women's education and the positions and activities of women in other countries. Náprstek allowed women to study in the drawing room and library of his house and organised lectures for women given by leading Czech philosophers and scientists. Among the most active ones was František Josef Studnička (1836–1903), who in 1865 opened his series of lectures for women with a paper on astronomy. Studnička had studied mathematics and physics at the University of Vienna; he was interested in natural science and geography. He was a secondary school teacher and later university professor in Prague. In his lectures, Studnička acquainted women with the latest news from astronomy and the tools and instruments used to observe the stars, he showed them the purpose of a barometer and the newest methods of time measurement. Lectures given by Josef Durdík (1837–1902), a professor at the university in Prague since 1880, were also very popular. In his dissertation, Durdík explored the philosophy of Leibniz and Newton. Later, he was influenced by the metaphysical and psychological opinions of Herbart. In his philosophical system, Durdík also incorporated the ideas of Charles Darwin, whom he had once met in person. Through his lectures, he taught Czech women about the key figures of the history of Czech philosophy and science, including Copernicus, Lavoisier, Bruno, Mill, Leibniz, and Kant, and about Darwin's evolutionary theory, ethics, and aesthetics ('On Character', 'On Nobility and Tragic Nature', 'On Beauty in Nature'). The American Ladies' Club was a free association. Its activities were supervised by the police and its existence was only officially permitted by the authorities in 1907. This might be the reason that its archives are not very extensive. The attendance sheets that have been preserved have the names of the lecturers and titles of the lectures (cf. Secká 2012). Lecturers included both men (Albín Bráf, Otokar Hostinský, Jan Evangelista Purkyně, Tomáš Masaryk, Jaroslav Vrchlický, and many others) and women. The most active was Sofie Podlipská, followed by Karolina Světlá, Věnceslava Lužická-Srbová, Ludmila Šimáčková, Klemeňa Hanušová, Anna Bayerová, and others. Before the first girls' grammar school was opened in 1890, the lectures given in the club may have been the main source of scientific information from different areas of philosophy and science for women and an impulse for the development of their scientific interests.

One of the club's members who carried out her own research work and published books on the history of astronomy was Paulina Šafaříková (1836–1920), née Králová. Šafaříková shared her interest in astronomy with her husband Vojtěch Šafařík (1829–1902). Šafařík had studied chemistry at universities in Prague, Berlin, and Göttingen. He contributed to the creation of Czech chemistry terms and was inspired by the Romantic approach to natural research (he translated Alexander von Humboldt's *Kosmos*). In 1868, he became a professor of chemistry at the polytechnic university in Prague; in 1882, he became a professor of astronomy at the Czech part of the university in Prague. He and his wife founded a private observatory in their house to make astronomical observations. In 1897, Šafaříková published her work *Z dějin dalekohledu* [*From the History of the Telescope*], introducing the key figures of optics and astronomy and presenting a detailed overview of the development of devices used to observe the sky and universe, from the remote past to the present. The work includes descriptions of the properties of light and information on the construction of individual devices, explanations of their functions, and other details. Šafaříková's work has the typical features of positivist interest in the history of science emanating from the idea of progress.

Her second published work, *William Herschel a jeho sestra Karolina* [*William Herschel and His Sister Caroline*] (1900), tells the story of two figures from the history of astronomy. The motivation for writing this book might have been a feeling of relatedness to Caroline Lucretia Herschel (1750–1848) and the fact that the Herschel family originally came from Moravia, which they had to leave due to their protestant faith in the 1600s. When describing the fates of Caroline and William,

Šafaříková parted from the memoirs and correspondence published in 1876. In her text, she emphasises Caroline's hard work, her close cooperation with her brother, her own successful observations (discoveries of planets and nebulae), and the awards she obtained from the scientific community. In 1888, Šafaříková suggested that Viennese astronomer Johann Palisa (1848–1925) name one of the asteroids discovered in that year in honour of Caroline. He agreed and the asteroid discovered on 31 October 1888 was called 281 Lucretia. Another of the many asteroids he discovered is named after Šafaříková herself (278 Paulina).

3.5 First University Graduates

The absence of a higher education system for girls in the Czech lands was one of the main obstacles preventing women from entering the world of science. Education for women in the first half of the 1800s was basically a luxury. Only rich families from the upper-class bourgeoisie could afford homeschooling, private schools, or studies in foreign boarding schools for their daughters. The situation started to change in the 1860s when secondary schools offering general education for girls were founded with the support of municipal councils; the first one was established in Prague in 1863. However, these schools were not intended to prepare girls for further studies at universities; they aimed to teach them the knowledge and skills that were expected of middle-class women. In 1868, the first attempt at founding a grammar school (in Czech, gymnázium) for girls failed due to the lack of interested female students or, more accurately, the lack of interest of their parents. Women's clubs, which were the most active force in the development of girls' education in the Czech lands in the 1870s and 1880s, focused on practical courses and on founding technical and commercial schools offering alternative options for widows or unmarried women who were financially dependent on their relatives. Girls who had predispositions for academic studies and longed for university education had to go abroad. They initially headed for Swiss universities. Despite various obstacles, particularly financial ones, two women managed to successfully complete their medical studies: Bohuslava Kecková and Anna Bayerová.[5] However, as the stories of the two first Czech female doctors show, it was not a very effective strategy for praxis. After they returned to Bohemia, the authorities refused to recognise their education and even rejected their requests to open a private practice. Neither woman was allowed to live up to her achieved level of education.

Bohuslava Kecková (1854–1911) grew up in Prague, where her father rented limestone quarries. In her memoirs, published in the form of a biographical sketch in *Ženské listy* in 1898, she mentioned she had been keen on studying from a very young age, and she had disliked doing housework (Kecková 1898, 66). Her father influenced her further academic orientation, encouraging her to pursue her interests.

[5]Based on contemporary reports and surviving correspondence, Julie Kurková is also said to have left for Zurich to study at university. She died of tuberculosis while studying at the Faculty of Arts.

After graduating from the Girls' High School, Kecková started to privately attend classes at a regular grammar school and gradually took various exams. Based on a special permit in 1874, she was allowed to take the graduation exam at a grammar school for boys. That autumn (in October 1874), she left for Zurich, accompanied by her father, where she enrolled in the medical faculty of the local university. She successfully completed her studies in 1880. In her dissertation thesis, she dealt with the problem of the enlarged thyroid. Upon her return to the Czech lands, where she learned that having her diploma recognised and starting a career in her field were not very likely, she decided to take a course in midwifery at the University of Vienna. From 1883, she worked in Prague as a midwife. Her further career was affected by the decision of the Austrian government in 1891 to create positions for female doctors for the then-occupied regions of Bosnia and Herzegovina. These posts were created as it was necessary to improve the health conditions of Muslim women, whose religion prevented them from being treated by men. Kecková worked as a state physician from 1893 until her death. She remained in touch with the Czech women's movement by mail. She published minor texts in women's magazines, in which she raised public awareness in the field of hygiene and health science; she also introduced the Czech public to customs and traditions in Bosnia and Herzegovina.

Anna Bayerová (1853–1924) came from a rural family and, unlike Kecková, her parents were rather conservative and initially showed no understanding for their daughter's studies. However, she finally managed to obtain their consent and left for Prague in 1868 to study at the Girls' High School. Later on, she privately prepared herself and took exams at a grammar school for boys. Bayerová became a member of the American Ladies' Club and attended its lectures, where she met Karolina Světlá, Sofie Podlipská, and Eliška Krásnohorská, who thought Bayerová was very gifted. Her decision to go abroad to study medicine at university may have been influenced by the contemporary role models and discussions about women's predispositions to be engaged in medicine. In several lectures for the American Ladies' Club, Ludmila Šimáčková, née Křížková (1844–1879) spoke about women's work in medicine (cf. Šimáčková 1872). Female medical students were also encouraged by Vítězslav Janovský (1847–1925), a doctor and later professor at the university in Prague (in 1874, he gave a lecture for the American Ladies' Club called O poměru ženy k lékařství ['On Woman's Attitude to Medicine']). In 1875, Bayerová went to Zurich where she applied for enrolment in university, promising to complete her secondary school studies later. Particularly for financial reasons, she switched after three terms to the University of Bern where it was cheaper to study and live. However, she continued to suffer from a lack of money, which even led her to interrupt her studies for some time. She returned to Prague where she was allowed to observe selected lectures at the Medical Faculty (cf. Honzáková 1937). Thanks to her friends' support, Bayerová finally returned to finish her degree studies in Bern, where she graduated in 1881. Her dissertation work, dealing with the blood of newborn babies, was entitled *Über die Zahlenverhältnisse der roten und weissen Zellen im Blute von Neugeborenen und Säuglingen* [*On the Proportion of Red and White Corpuscles in the Blood of Newborns and Infants*]. Bayerová worked in sanatoriums and hospitals in many different places in Germany and Switzerland (e.g. Dresden, Baden-Baden,

Leipzig, Bern, and Teufen); she also spent one year (1892) as a state physician in Bosnia but she longed to return to the Czech lands. She was enabled to do so in 1896, when a regulation was issued allowing the recognition of medical degrees awarded abroad. Nonetheless, the conditions determined by this regulation were unacceptable for Bayerová. If she wanted to work as a physician in the Czech lands, she would have had to retake exams, including the secondary school leaving exam. Hence she went abroad for one more time and returned to Prague for good when she was 56. She then earned her living as a private nurse and teacher. The Austrian authorities did not let her treat people until the outbreak of the First World War. During her studies, Bayerová started to contribute to women's magazines. Her reports about contemporary methods of blood transfusion and the possibilities of infant formulas were published in the *Časopis lékařů českých*, one of the oldest Czech journals (published from 1862). Bayerová also translated into Czech the popular and widely read work *Die Frau als Hausärztin* [*The Woman as a Family Doctor*] (1901) by Anna Fischer-Dückelmann (1856–1917). The Czech translation was published as *Žena lékařkou* in 1920.

The dream of a secondary school for girls offering the same education as a school for boys and enabling its female graduates to continue studying at university came true in the 1890s, particularly thanks to the endeavours of Eliška Krásnohorská, who managed to find support for this project from a number of prominent and influential figures. A detailed account of the foundation and history of the first grammar school for girls, named Minerva after the Roman goddess of wisdom, is given in the annual report and chronicle of this school (cf. Sekyrková 2016). The school was authorised in 1890 as a private educational institute, which had a major impact not only on the financial aspect and accessibility of studying there, but also on the form of final exams. The Austrian government did not support the school financially, it did not incorporate it into the system of government-subsidised schools, and its students had to apply every year to be allowed to take the final graduation exam at a grammar school for boys administered by the state.[6] It was more demanding to study there than at a boys' grammar school, as girls had to master the same curriculum in five years as boys had to learn in eight years. Despite that, many girls wanted to study there. In its first year, over fifty girls commenced their studies at Minerva, and the number of girls attending the school was more or less constant for the next seven years. The number of students doubled in the school year 1897/1898, when it became clear that Minerva graduates would be able to study full-time at the Faculties of Arts of Austrian universities.

After Minerva was founded, there were efforts to give the graduates of a grammar school for girls access to university studies. In 1890, *Ženské listy* published a petition addressed to the Imperial Council requesting that girls be allowed to study at secondary schools and universities, just as they were in other developed countries. The petition was submitted by the MP Karel Adámek (1840–1918), who was a member of the Národní strana svobodomyslná [National Liberal Party],

[6] Minerva obtained the so-called right of the public, which authorised the school to administer graduation exams, in the school year 1907/1908.

promoting national and liberal ideas. The presented arguments did not persuade other lawmakers. The first grammar school for girls had to be run as a private institute, and its graduates had to wait until 1897 for official permission to study at university. However, the atmosphere at the university in Prague was favourable towards supporting girls' education. At that time, many professors, particularly at the Czech Faculty of Arts,[7] supported women in gaining access to university studies. The surviving *Chronicle* of Minerva shows that female students went on excursions to various university premises and departments (the department of physics and geography, the observatory, the university library, etc.) where they were introduced to exhibits and instruments, and natural phenomena and laws were demonstrated and explained to them. In early 1895, when the first Minerva graduates were completing their studies, the school management turned to the rector of the university in Prague and asked whether their students could attend lectures as observers. In exceptional cases, women were allowed to attend university lectures by an 1878 decree; they could observe lectures only with the permission of the lecturer teaching the particular subject. According to the *Chronicle*, nearly all staff members of the Faculty of Arts of the university in Prague agreed; only three refused (Sekyrková 2016, 58). Degree and non-degree study was available to women at the Faculties of Arts of the Imperial and Royal universities (in the Austrian part of Austria-Hungary) in 1897. That year, 10 degree and 16 non-degree female students were admitted to the Faculty of Arts of the Czech university (Havránek 1997, 193). The lectures that Minerva graduates had attended in the previous years as observers were recognised; therefore, the first women graduated from university in 1900. By the end of 1900, the following women had passed the final state exam at the Czech Faculty of Arts: Jindřiška Hrabětová, Božena Jiránková, Marie Fabiánová, Olga Šrámková, and Helena Tuskányová. The number of women at the Czech university in Prague gradually increased until 1918. Women initially comprised approximately 4 to 10% of the total number of students;[8] the proportion of female students went up to 30 to 40% between 1914 and 1918 (cf. Havránek 1997, 355). The graduates of the Faculty of Arts worked especially as teachers at secondary schools for girls, which were being founded in all of the larger cities of Bohemia, Moravia, and Silesia.

In addition to teacher and governess, another job regarded as appropriate for women in the society of that time was nurse and caregiver. When the first Minerva graduates were striving to enter universities, at least as observers, there was an ongoing debate in the Czech press on whether it was appropriate for women to study medicine at universities. One of the main stimuli of this debate was the text *Die Frauen und das Studium der Medicin* ['Women and Medical Studies'] (1895) by the Vienna-based professor Eduard Albert (1841–1900), who criticised women studying medicine. Albert substantiated his argument specifically by pointing out the size of women's brains (women have smaller brains), which he associated with insufficient intellectual capabilities, and noting the physical and mental demands of

[7]The university in Prague was divided into Czech and German parts in 1882.

[8]The highest number of students studied at the Faculty of Law; this Faculty did not admit any women at all.

medical studies and the medical practice itself. Even though it cannot be said there were only critics of women's medical studies among the staff of the university in Prague, the fact is that the Czech Faculty of Medicine did not admit any women for non-degree study when it was approached in 1895 by the management of the grammar school for girls. Hence Minerva students decided to try their luck at the German Faculty of Medicine, which complied with their request. In 1896, three Minerva students enrolled in the German Faculty of Medicine: Anna Honzáková, Rosalie Machová, and Marie Peigerová. They transferred to the Czech Faculty of Medicine two years later in 1898 due to the national unrest at the end of 1897, which made it impossible for them to stay in the German part of the university. Women's degree and non-degree study at medical faculties of Imperial and Royal universities was allowed in 1900. Female medical students could start right away; applicants for pharmaceutical studies were admitted starting from the school year 1903/1904 (cf. Štemberková 1996, 214). The first female Doctor of Medicine graduating from the Czech Faculty of Medicine was Anna Honzáková (1875–1940), who graduated in 1902. By the end of 1918, 60 more women had graduated from the Faculty of Medicine of the Czech university in Prague. The first woman to graduate with a degree in pharmaceutical studies was Růžena Kroutilová (1884–1974) in 1909.

Initially, the alumnae of medical faculties specialised particularly in gynaecology and paediatrics; however, they gradually extended their specialisations to other medical branches as well. Nonetheless, their chances of finding a job were slim. In the beginning, female doctors found work only in clinics led by chief physicians supporting female doctors. Another option was opening and running a private practice. A majority of the first graduates of medicine were involved in raising public awareness and published texts about hygiene and health science in women's magazines. Until 1918, their articles occurred in specialised medical journals only rarely. For example, two texts by Marie Peigerová (1877–1967) were published in 1905 in the *Časopis lékařů českých*, entitled *Nová pozorování o substituci svalové* ['New Observations on Muscular Replacement'] and *Spontánní výbuchy smíchu při hemiplegii* ['Spontaneous Outbursts of Laughter in Hemiplegia']. Peigerová graduated from the Czech Faculty of Medicine in 1904 and later worked in the clinic of internal medicine led by chief physician Professor Josef Thomayer (1853–1927). The dedications of the two aforementioned articles indicate that Thomayer supported and supervised Peigerová in writing the texts. In 1910, Peigerová moved to Brno, where she ran a private practice and worked as a physician at schools for girls (cf. Navrátil 1913).

Studies at the Faculty of Arts of the university in Prague could be completed in two ways. Most students completed their studies by passing a state exam before a committee for grammar schools, which qualified successful graduates to teach at secondary schools. Another option was to write and defend a dissertation thesis and pass two doctoral (*rigorosum*) examinations. In that case, the graduate earned the degree of a Doctor of Philosophy, which was an essential prerequisite for further work at university and for a career as a researcher. The first female graduates of the Faculty of Arts had only a slight chance to have an academic career; despite that, many of them tried to obtain a doctoral degree and some eventually succeeded. In 1901,

Marie Zdeňka Baborová (1877–1937) received a doctorate in zoology and Marie Fabiánová (1872–1943) in mathematics (cf. Bečvářová 2019). By 1918, another 42 women had obtained a doctoral degree at the Czech Faculty of Arts. As regards the topics of their dissertation works, ten dealt with natural scientific topics, eight with history, five dissertations pertained to classical philology, five were in Czech studies, four in philosophy, three explored a mathematical-physical topic, three were in Romance studies, two in geography, and two in German studies (Štemberková 1996, 216). When inspecting the registers of university in Prague, attention is drawn to the name of Klára Kohoutová (1882–?), who graduated from the Faculty of Arts in 1912 (dissertation work: 'Can Schopenhauer's Pessimism Have an Educational Function?') and from the Faculty of Medicine in 1920. Besides teaching, which was the most frequent occupation, female graduates worked in libraries or newspapers; they also devoted a lot of their energy to raising public awareness and to emancipation activities. Attempts at their own research work were complicated; some of them will be dealt with in the next chapter in greater detail. The lives of the first female doctors of philosophy have been largely forgotten.

An example of a great talent and gift that has disappeared from the Czech cultural heritage is the dissertation work *Pojem apercepce u Herbarta a Wundta buď vyložen, srovnán a posouzen* ['The Notion of Apperception in Herbart and Wundt, Explained, Compared, and Evaluated'] by Emilie Veselá (1874–?). University registers show that Veselá entered the Faculty of Arts in 1901 and earned her degree in 1911. Her name can be found in the Minerva *Chronicle*, where she was mentioned as the record-keeper of the association and later, after graduating from university, as a teacher. In her work, Veselá tried to compare and evaluate how apperception was understood by Johann Friedrich Herbart (1776–1841) and Wilhelm Wundt (1832–1920), who represented two influential and, in many respects, contradictory conceptions of the late 1800s. In her critique, Veselá was influenced by the ideas of her teacher František Krejčí (1858–1934), who attempted to build psychology upon purely empirical grounds. Veselá assessed Herbart's and Wundt's conceptions of apperception, particularly according to how much it was possible to eliminate metaphysical and speculative elements and the degree to which they were in line with the scientific findings of that time. Veselá noted that Wundt went beyond Herbart's psychological mechanisms and extended mental processes (in addition to cognitive features) to include volition and emotion. However, she saw it as problematic that Wundt, when explaining the process of apperception, placed a great emphasis on will, which he perceived in a very broad sense. According to Veselá, Wundt's conception of apperception contravened experience as well as some findings of modern physiology and psychiatry. Although Herbart's conception of apperception was based on his metaphysics, Veselá believed it was possible to eliminate these speculative foundations and bring it into line with modern scientific knowledge. Hence, she concluded that Herbart's conception of apperception was more significant and fruitful for the development of empirical psychology.

Women's Rights and Animal Rights

Two of the prime objectives of Czech feminists in the 1900s were women's suffrage and active participation in politics. Women started to put forward their demands in the realms of civil and political rights in the 1890s, a period when the Czech women's movement internally diversified. The new generation of women entering the scene at that time were able to benefit from their achievements in education and started to make more claims that would lead to gender equality in all spheres of public life. Women's participation in political activities was not acknowledged in the legal system of that time (the 1867 'Association Law' banned women's membership in political organisations); despite that, women attended political rallies and meetings and were also represented in the leaderships of some party organisations. In 1905, there were demonstrations and strikes for universal and equal suffrage in the Czech lands. Even though none of the political parties explicitly promoted women's right to vote, the activists of the contemporary women's movement embarked on an independent path, founding the Committee for Women's Suffrage in late 1905 (cf. Kořalka 1996, 309). Despite all the efforts, women failed to achieve their goal. A 1907 reform abolished the five-class suffrage system: only men over 24 years were entitled to vote. Nevertheless, the activities of the Committee for Women's Suffrage had many positive outcomes. Women won the support of other influential figures and managed to become more closely integrated in the structures of political parties.

In advancing their interests, Czech women used quite moderate methods (raising public awareness and petitions), benefited from the complicated character of the legal system, and made references to formalist interpretations of the existing laws. For example, in the debate on passive suffrage, they pointed out that the rules for the elections for the Czech Diet from 1861 did not explicitly exclude women from standing in elections.[9] They drew on that in the 1908 elections. The Social Democrats nominated Karla Máchová (1853–1920), a teacher and editor who had spent several years in the United States in the 1890s working in the compatriot women's movement. Marie Tůmová (1867–1925), a teacher at a Volksschule for girls, stood as an independent candidate, supported by the Committee for Women's Suffrage. National parties also agreed on a joint female candidate, nominating Božena Zelinková. Even though none of these women were elected, it was evident from the results that the inhabitants of the Czech Kingdom were at least changing their attitudes towards women's participation in politics. Women celebrated their first win in the elections four years later, in the 1912 by-elections for the Czech Diet, due to the death of an MP. At that time, there were two women among the candidates, Karla Máchová and Božena Viková-Kunětická, née Novotná (1862–1934), who eventually won in the second round. The fact that the popular mood was changing is evident from the fact that Franz, Prince of Thun, the Imperial-Royal governor of Bohemia, had already notified the Austrian Ministry of the Interior before the by-elections that a woman was likely to be elected and had asked them to release a statement on whether such a result would be recognised. He himself believed that, based on the

[9] Electoral codes for the Moravian Diet and Imperial Council explicitly excluded women from active and passive voting rights.

contemporary rules, women did not have passive voting rights for the Diet. After being elected, Viková-Kunětická was congratulated by representatives of Czech and foreign women's movements, yet she was never issued an official certificate to be able to enter the Diet (cf. Kořalka 1996, 314–320).

When exploring the history of the women's movement, it becomes clear that in some countries the fight for women's rights was closely tied with animal rights.[10] This was evident in the controversy over vivisection that took place in the late 1800s and early 1900s. Many women became involved in discussions of whether it was justifiable to use animals for scientific purposes, siding with anti-vivisectionists who demanded a complete ban on experimentation on live animals. The intersection and interconnection of anti-vivisectionists and the women's movement was especially notable in England, where the dispute started and significantly entered the public space (cf. Elston 1990; French 1975). However, the situation in the Czech lands was different; the issue of vivisection was quite marginal at first. No anti-vivisection movement was formed before the end of the 1800s, even though the dispute was thoroughly reported from the beginning. The request to impose a ban on vivisection did not appear until the early 1900s; nevertheless, an influential number of Czech feminists sided with those who defended experiments in medicine. The debate is also important from today's perspective, as it contributes to the understanding of women's attitudes towards science and scientific knowledge, which was strongly developed in the second half of the 1800s. Czech women's attitudes towards the vivisection issue was related to their conception of the importance and content of education for girls.

Campaigns against using animals for scientific purposes started in the Czech lands in the 1900s, i.e. at the time of the second anti-vivisection wave in England, which was provoked by the publication of the book *The Shambles of Science* in 1903 (cf. Lansbury 1985). Among Czech women, this topic was explored especially by Pavla Moudrá (1861–1940), a journalist, writer, and translator. Moudrá was very well acquainted with the contemporary debate on animal rights; she had translated and published the anthology by Henry Salt *Animals' Rights* (published in 1915 as *Práva nižších* [*Rights of the Inferior*]). She knew and drew on the major texts of anti-vivisectionists. Her attitude towards animals was also influenced by the ideas of Leo Nikolayevich Tolstoy and Arthur Schopenhauer. Her key text on this topic was the book *Vivisekce: úvahy o její ceně a prospěchu* [*Vivisection: Ideas about Its Cost and Benefit*] (1909), in which she introduced vivisection as a widespread abuse of animals, hidden from the public eye, that the state privileged and protected either by legally enabling it or by supporting the creation of physiological, pharmacological, and other institutes and laboratories where the experiments were conducted. She illustrated the gravity of the problem on the ever-increasing number of performed vivisections and used England to illustrate that this increasing number can still be observed where there is proper legislation (*An Act to Amend the Law Relating to Cruelty to Animals* 1876). Moudrá used the incapacity of the law to restrict the

---

[10]The parallel between speciesism and sexism is pointed out e.g. by Peter Singer in his work *Animal Liberation* (1975), which he illustrates with some of the same members of the animal rights movements and women's rights movements in the history of England and the United States.

number of experiments to justify her opinion that the only way to minimise animal suffering was to completely ban vivisection. Moudrá repeated the objections that had been used by anti-vivisectionists in the debate especially in the 1870s and 1880s. She claimed that the experimental method was neither important nor necessary for knowledge and progress in medicine. She played down the results that were obtained by means of experiments on animals and pointed out that this method often resulted in mistakes.

In her work, Moudrá used the same methods and strategy as the texts by Frances Power Cobbe (1822–1904), who is regarded as the most prominent female figure of the anti-vivisection movement in England. Hamilton calls this strategy, relying on the repeated use of selected sources, such as quotes from testimonies before the royal committee, descriptions of selected experiments, illustrations from physiological manuals, etc., *assemblage* (Hamilton 2010). One aim of this method was to make a complicated scientific issue more comprehensible for the general public and thereby reach and maintain an audience for anti-vivisection ideas. Hamilton argues that even though this strategy was not successful in drawing the public's attention, it was essential in establishing the identity of the anti-vivisection movement. Although Moudrá used the same methods in her work *Vivisection: Ideas about Its Cost and Benefit*, she failed to create a coherent movement that would adopt her attitudes and spread them further in an organised manner. On the contrary, an influential number of Czech feminists rejected anti-vivisection arguments and sided with those who supported experiments on animals in science.

An indirect criticism of Moudrá's attitude was expressed by Eliška Vozábová (1874–1973), one of the first female doctors to graduate from the university in Prague (she graduated in 1904), who earned considerable respect among Czech feminists. Her work *O vivisekci* [*On Vivisection*] (1910) was written at the initiative of the editorial board of *Ženský obzor* [*Women's Horizon*], which was one of the most progressive Czech women's magazines of that time. Vozábová's text was a clear defence of the use of experiments in medical science. Vozábová focused on the importance of vivisections for medicine as a theoretical discipline as well as the applicability of its results in practice. She introduced experiments on live animals to her readers as an indispensable research method resulting in findings that were central for determining the causes of illnesses and their efficient treatment. She saw imposing a ban on vivisections as 'ignorant' and, in the context of situations taking place at that time, she criticised the treatment methods of natural doctors and those who questioned the usefulness of vaccination.

It is possible to say, with a certain degree of simplification, that Moudrá did not succeed in achieving her goal of drawing Czech women's attention to the problem of animal abuse in science particularly because she put too much focus on trying to prove that vivisections were not necessary or beneficial. Claims of the uselessness and needlessness of experimental methods in medicine did occur in the texts of anti-vivisection societies, but they were quickly abandoned in favour of calls for a ban on experiments on live animals depicting cruelties that, according to critics, proved the amorality of this practice (cf. e.g. Cobbe 1892). To evaluate and comprehend the form of the controversy over vivisection in the Czech context, it must be emphasised that at the time when Moudrá published her text, the evidence for the claims of the beneficial effect of the experimental method in medicine was undergoing radical changes. The relativisation of the beneficial effects of the results of medical research may have been persuasive for some in the 1870s, but this line of argumentation ceased to be efficient in the early 1900s. Physiologists and doctors did not have to promise abstract future benefits; they could refer to specific achievements, including in the realm of public health. Hence, Moudrá's objections were unconvincing and unacceptable for women who strove to give girls access to education based on the results of science (cf. Jastrzembská 2017).

References

Bahenská M (2005) Počátky emancipace žen v Čechách: dívčí vzdělávání a ženské spolky v Praze v 19. století. Libri, Praha

Bahenská M, Heczková L, Musilová D (eds) (2010) Ženy na stráž!: české feministické myšlení 19. a 20. století. Masarykův ústav a Archiv Akademie věd České republiky, Praha

Bahenská M, Musilová D, Heczková L (2011) Iluze spásy: české feministické myšlení 19. a 20. století. Veduta, České Budějovice

Bečvářová M (2019) Doktorky matematiky na univerzitách v Praze 1900–1945. Karolinum, Praha

Bolzano B ([1810] 2007) On the mission and dignity of womanhood. In: Bolzano B (ed) Selected writings on ethics and politics. Rodopi, Amsterdam, pp 133–139

Bolzano B ([1827] 2009) Athanasia, neboli, Důvody pro nesmrtelnost duše. Stefanos, Jindřichův Hradec

Bolzano B ([1836] 1981) Vlastní životopis. Odeon, Praha

Bolzano B ([1847] 1951) O pokroku a dobročinnosti. Praha: Vyšehrad

Bolzano B ([1932] 2007) On the best state. In: Bolzano B (ed) Selected writings on ethics and politics. Rodopi, Amsterdam, pp 233–356

Bolzano B (2007) Selected writings on ethics and politics. Rodopi, Amsterdam

Čelakovský FL, Rajská B (1872) Z let probuzení. Kniha 2., Listy Fr. Lad. Čelakovského a Bohuslavy Rajské 1844–1845. J. Otto, Praha

Červinková M (1881) Bernard Bolzano: životopisný nástin. M. Červinková-Riegrová, Praha

Červinková M (1887) Ochrana chudé a opuštěné mládeže: rozhledy po lidumilství v Evropě. Nákladem spolku Ochrana opuštěných a zanedbaných dívek v komisi universitního knihkupectví Bursíka a Kohouta, Praha

Červinková M (1892) Marie Riegrová, rodem Palacká: její život a skutky. Národní tiskárna a nakladatelství, Praha

Červinková M (2009) Zápisky I (1880–1884). Národní archiv, Praha

Červinková M (2013) Zápisky II (1885–1886). Národní archiv, Praha

Cibulka P, Hájek J, Kučera M (2009) The definition of Czech National Society during the period of liberalism and nationalism (1860–1914). In: Pánek J, Tůma O et al (eds) A history of the Czech lands. Karolinum, Prague, pp 329–376

Cobbe FP (1892) The nine circles of the hell of the innocent: described from the reports of the presiding spirits. Swan Sonnenschein and Co., London

Elston MA (1990) Women and anti-vivisection in Victorian England, 1870–1900. In: Rupke N (ed) Vivisection in historical perspective. Routledge, London, pp 259–294

Faktorová V (2012) Poetika obrozenské vědy a pluralita romantismu. In: Tureček D et al (eds) České literární romantično. Host, Brno, pp 239–262

French RD (1975) Antivivisection and medical science in Victorian Society. Princeton University Press, Princeton, N.J.

Gascoigne J (2003) Ideas of nature: natural philosophy. In: Porter R (ed) The Cambridge history of science. Cambridge University Press, Cambridge, pp 285–304
Hájek J, Hlavačka M (2009) The birth of the modern Czech Nations (1792–1848). In: Pánek J, Tůma O et al (eds) A history of the Czech lands. Karolinum, Prague, pp 281–309
Hamilton S (2010) Reading and the popular critique of science in the Victorian anti-vivisection press: Frances Power Cobbe's writing for the Victoria Street Society. Vic Rev 36:66–79
Haubelt J (1996) Alexander von Humboldt a vzdělanost ve středu Evropy. In: Blucha J (ed) Alexander von Humboldt a České země. Humboldt klub České republiky, Praha, pp 38–55
Havránek J (ed) (1997) Dějiny Univerzity Karlovy. III, 1802–1918. Karolinum, Praha
Hlavačka M (2009) Czechs during the revolution and neo-absolutism (1848–1860). In: Pánek J, Tůma O et al (eds) A history of the Czech lands. Karolinum, Prague, pp 311–328
Hoffmann J ([1850] 2016) Bruchstücke zu einer künftigen Lebensbeschreibung des Bernard Bolzano. In: Hoffmann J et al (eds) Beiträge zu Bolzanos Biographie. Frommann-Holzboog, Stuttgart-Bad Cannstatt, pp 15–135
Hoffmannová E (1982) Karel Slavoj Amerling. Melantrich, Praha
Honzáková A (1937) Anna Bayerová 1853–1925: první česká lékařka ve Švýcarech. Ženská národní rada, Praha
Iliffe R (2003) Science and voyages of discovery. In: Porter R (ed) The Cambridge history of science. Cambridge University Press, Cambridge, pp 618–646
Jahn JV (1893) Karel Slavoj Amerling: obraz života a práce. F. Šimáček, Praha
Janko J, Štrbáňová S (1988) Věda Purkyňovy doby. Academia, Praha
Jastrzembská Z (2017) Spor o vivisekce a české ženy na přelomu 19. a 20. století. Teorie vědy 39:187–205
Klecková B (1898) MUDr. Bohuslava Kecková, bosensko-hercegovská zemská lékařka v Mostaru. Ženské listy 26:66–70, 90–93
Klášterský I, Hrabětová A, Duda J (1982) Dějiny floristického výzkumu v Čechách, na Moravě a ve Slezku. I. Okresní vlastivědné muzeum v Litoměřicích, Litoměřice
Kořalka J (1996) Zvolení ženy do českého zemského sněmu roku 1912. In: Pešek J, Ledvinka V (eds) Žena v dějinách Prahy. Scriptorium, Praha, pp 307–320
Kostlán A, Niklíček L (2018) Scholarly and scientific institutions and associations in the Czech lands from the end of the 18th century to 1882. In: Míšková A, Franc M, Kostlán A (eds) Bohemia docta: the historical roots of science and scholarship in the Czech lands. Academia, Praha, pp 149–199
Kruta V (1971) J.E. Purkyně's contribution to the cell theory. Clio Med 6:109–120
Kučerová V (1914) K historii ženského hnutí v Čechách: Amerlingova éra. Ženská revue, Brno
Lansbury C (1985) The old brown dog: women, workers, and vivisection in Edwardian England. University of Wisconsin Press, Madison
Loužil J (1971) Ignác Jan Hanuš. Melantrich, Praha
Loužil J (1978) Bernard Bolzano. Melantrich, Praha
Loužil J (1987) Jan Evangelista Purkyně - přírodovědec a naturfilozof. In: Purkyně JE (ed) Útržky ze zápisníku zemřelého přírodovědce. Mladá fronta, Praha, pp 51–72
Macura V (1998) Příběh encyklopedie dam. Tvar 5:4–5
Macura V (1999) Salon u Fričů a konverzační slovník pro dámy. In: Lorenzová H, Petrasová T (eds) Salony v české kultuře 19. století. Koniasch Latin Press, Praha, pp 75–87
Matoušek O (1954) Lékaři a přírodovědci doby Purkyňovy: životopisné studie. Státní zdravotnické nakladatelství, Praha
Moudrá P (1909) Vivisekce: úvahy o její ceně a prospěchu. Hejda and Tuček, Praha
Navrátil M (1913) Almanach českých lékařů: s podobiznami a 1000 životopisy: na paměť 50 letého jubilea Spolku a Časopisu lékařů českých. Nákladem spisovatelovým, Praha
Němcová B (1955) Národopisné a cestopisné obrazy ze Slovenska. Státní nakladatelství krásné literatury, Praha
Němcová B (1957) Národopisné a cestopisné obrazy. Státní nakladatelství krásné literatury, Praha

Němcová B (2004) Božena Němcová – Korespondence II (1853–1856). Nakladatelství Lidové noviny, Praha
Neudorflová ML (1999) České ženy v 19. století: úsilí a sny, úspěchy i zklamání na cestě k emancipaci. Janua, Praha
Ottův slovník naučný (1888–1909) J. Otto, Praha
Pánek J (1974) Pavla Moudrá. Poznámky k životu, působení a pozůstalosti české spisovatelky a bojovnice za mír. Středočeský sborník historický 9:215–248
Paulová N (1862) Vítězka Paulova. Životopisný nástin. Lada 2:12–13
Paulová V (1861) Věk člověka. Lada 1:28–29, 41–42, 51–52, 67–68, 76–77, 92–93
Pavlíková M (1985) Bolzanovo působení na pražské univerzitě. Univerzita Karlova, Praha
Pluskal FS (1849) Biographie der berühmten, jetzt lebenden Pflanzenforscherin Oesterreich's, Frau Josephine Kablik. Franz Gastl, Brünn
Pokorná M (2018) Royal Bohemia Society of Sciences. In: Míšková A, Franc M, Kostlán A (eds) Bohemia docta: the historical roots of science and scholarship in the Czech lands. Academia, Praha, pp 84–148
Purkyně JE ([1819] 1969) Příspěvky k poznání zraku ze subjektivního hlediska. Univerzita J. E. Purkyně, Brno
Purkyně JE ([1850] 1987) Útržky ze zápisníku zemřelého přírodovědce. Mladá fronta, Praha
Purkyně JE (1857) O vzniknutí časopisu Kroku i zaniknutí jeho. Živa 1:83–86
Rajská B, Němcová B (1873). Z let probuzení. Kniha 3., Vzájemné dopisy Antonie Čelakovské (Bohuslavy Rajské) a Boženy Němcové 1844–1849. J. Otto, Praha
Rajská B (1872) Z let probuzení. Kniha 1., Paměti a korespondence Bohuslavy Rajské z let 1839–1844. J. Otto, Praha
Rettigová MD (1840) Mladá hospodyňka w domácnosti, jak sobě počínati má, aby své i manželovy spokojenosti došla: dárek dcerkám českoslovanským. Jan H. Pospíšil, Praha
Rudová J ([1863] 1876) Loterie, hrst naděje a pytel nejistoty, neboli, Co jest loterie?: máme-li v ní sázeti?: jakého štěstí můžeme se z hry té nadíti?: s velkými početními tabulemi. Urbánek, Praha
Rudová J (1864) Tabák kuřlavý a šňupavý, kouření a šňupání: knížka jak šňupákům a kuřákům potřebná, tak nekuřákům a nešňupákům užitečná. Josef Ruda, Praha
Šafaříková P (1897) Dějiny dalekohledu. J. Otto, Praha
Šafaříková P (1900) William Herschel a jeho sestra Karolina: biografický nástin. P. Šafaříková, Praha
Secká M (2012) Americký klub dam: krůček k ženské vzdělanosti. Národní muzeum, Praha
Sekyrková M (1990) K počátkům Komitétu pro přírodovědný výzkum Čech. Dějiny věd a techniky 23:212–220
Sekyrková M (2016) Minerva 1890–1936: kronika prvního dívčího gymnázia v habsburské monarchii. Univerzita Karlova, nakladatelství Karolinum, Praha
Šimáčková L (1872) Vynikající ženy mimo rodinný kruh. Libuše, Matice zábavy a vědění. F. Šimáček, Praha
Štemberková M (1996) Doktorky filozofie a medicíny na pražské univerzitě od r. 1901 do konce první světové války. In: Pešek J, Ledvinka V (eds) Žena v dějinách Prahy. Scriptorium, Praha, pp 213–234
Tretera I (1989) J.F. Herbart a jeho stoupenci na pražské univerzitě. Univerzita Karlova, Praha
Urválková Z (2009) Dvojlomná zrcadlení: dílo Karla Herloše-Herloßsohna v českém literárním kontextu. ARSCI, Praha
Vozábová E (1910) O vivisekci. In: Vozábová E (ed) O vivisekci. Vliv alkoholu na život ženy a dítěte. Ženský obzor, Praha, pp 1–18
Yeo R (2003) Classifying the sciences. In: Porter R (ed) The Cambridge history of science. Cambridge University Press, Cambridge, pp 239–266
Zapová H (1863) Nezabudky: dar našim pannám. Kober, Praha
Zouhar J (2008) Dějiny českého filozofického myšlení do roku 1968: stručný přehled. Academicus, Brno

Chapter 4
Czech Women in Philosophy and Science Since 1918

Abstract After the 1918 foundation of Czechoslovakia, more opportunities to study at universities opened up for women. Women were gradually admitted to all faculties, except for theological faculties, at universities and colleges and started to work in various fields of philosophy and science, following their own professional interests. The period until 1939 shows women's efforts to get involved in scientific structures, achieve a certain form of recognition, and participate in educating the next generation of researchers. Hence, women's relationships to official scientific institutions are used here as a classification criterion for the given period. One of the most interesting discussions, one which had a distinct philosophical dimension and affected ideas about the further orientation of the women's movement, pertained to the issue of abortion and a woman's right to have control over her body. During the Second World War, Czech universities were closed by the Nazi occupants, many Czech scholars were persecuted, and research was limited to private activities. After 1948, Marxism became the only acceptable opinion in the humanities; other approaches were not acknowledged. The presentation of women in Czech science and philosophy between 1945 and 1989 starts by explaining the specific status of women in the socialist order and shows the types of Czech women in science and philosophy between 1945 and 1989, based on their ranking in academic institutions and their relation to the socialist regime.

4.1 In the Spirit of the First Republic

4.1.1 Historical and Social Context

In terms of women's rights, the foundation of the first independent Czechoslovak state, declared on 28 October 1918, was an important milestone. The principle of gender equality in all spheres of life had already been explicitly stated in the 'Declaration of Independence of the Czechoslovak Nation by Its Provisional Government' (the

Z. Jastrzembská et al., *Czech Women Philosophers and Scientists*,
SpringerBriefs in History of Science and Technology,
https://doi.org/10.1007/978-3-030-61516-1_4

so-called Washington Declaration, 18 October 1918). The authors of this document, aiming at formulating the main principles of the constitution of the Czechoslovak nation, emphasised freedom of speech, religion, and scientific research, and that 'women shall be placed on equal footing with men, politically, socially, and culturally'. This idea was then embedded in the Constitution of the Czechoslovak Republic, adopted in 1920 (§106 'Privileges due to sex, birth, or occupation shall not be recognised.').

Immediately after the foundation of the Czechoslovak Republic, opportunities for women to study at university expanded. In addition to faculties of arts and medicine, women were allowed to study at the faculty of law, which was one of the largest faculties at the university in Prague in terms of the number of students. In 1918, 80 women enrolled in the Faculty of Law at the Czech university in Prague. In the interwar period, women comprised approximately 10% of the total number of students at this faculty (cf. Bečvář et al. 1998, 598–602). Towards the end of 1922, the first woman to become a Doctor of Law was Anděla Jírová, née Kozáková (1897–1986). Six years later (1928), she passed the notary exam; she started working as a notary ten years later. Women's studies became more common at the newly founded universities as well. In 1919, Masaryk University in Brno and Comenius University in Bratislava were established. Female applicants were also admitted to technical colleges. Having succeeded in gaining access to education and having obtained a passive and active right to vote, the feminist movement in the newly founded Czechoslovakia focused on the issue of women's employment and careers. In order to accomplish the process of transferring the declared ideas of gender equality into social practices and everyday life, the Women's National Council was established in 1923. This organisation united and coordinated activities particularly of those women's associations that were close to the liberal-progressive movement.[1] Liberalism has been one of the main ideological sources of women's emancipation since the 1800s. In Czech society, its principles were developed, though at the same time criticised, by Tomáš Garrigue Masaryk (1850–1937), who became the first Czechoslovak president. To some female representatives of the women's movement, Masaryk embodied the support of their endeavour and promise of their future success. In 1930, to celebrate Masaryk's 80th birthday, the Women's National Council published an anthology entitled *Masaryk a ženy* [*Masaryk and Women*]. The authors collected and printed Masaryk's lectures and texts related to women's emancipation and added their personal thoughts and memories. One of the main initiators of this anthology was Františka Plamínková (1875–1942), a politician, journalist, leading figure, and long-time chair of the Women's National Council. Under her leadership, the Council's activities focused mainly on commenting on laws proposed in the parliament, analysing the existing legislature, and preparing individual proposals of changes that the liberal-oriented part of the women's rights movement regarded as most essential. The organisation also joined the international structure and became part of the International Women's Council,

[1] In the interwar period, along with the liberal-oriented Women's National Council, there were Catholic women's associations, Christian-social associations, and working women's associations. The German women's associations in Czechoslovakia had their own structure.

founded in 1888 in Washington, D.C. Due to external circumstances and growing pressure from German Nazi occupiers, the Women's National Council dissolved itself in 1942.

One of the most interesting discussions, having a distinct philosophical dimension and affecting the orientation of the women's movement in the 1920s and 1930s, pertained to the issue of abortion and a woman's right to control her own body. In the Czechoslovak Parliament, three bills were proposed within a mere five years to annul the then-valid legislation of the 'crime of foetus expulsion' and to make abortions legal under certain conditions. The perception and punishment of abortion as a crime was in accordance with §§144–148 of the imperial penal code n. 117/1852, which applied to the entire Czech territory in order to maintain legal certainty and continuity after the foundation of the independent Czechoslovak state. According to this norm, women faced up to five years in prison for having an abortion. The same punishment applied to those who assisted the women. The first proposal of an amendment attempting to annul the relevant paragraphs of the imperial law, based on the belief that women have the right to freely control their own bodies, was proposed to the parliament in 1920. The bill was formulated by MPs from the socialist party, one of whom was Luisa Landová-Štychová, née Vorlíčková (1885–1969). For her radical ideas and tendencies towards communism and anarchism, Landová-Štychová is usually regarded as a pioneer of 'anarcho-feminism'. In particular, the advocates of the amendment accentuated the harsh economic conditions that usually affect a woman's decision to have an abortion, and the problematic nature of the practice that incarcerates women who are often acting in order to improve the life situation of the children they already have. In the text of the bill, the authors strongly emphasise the autonomy of the decision and the health aspect of abortions. They assert that abortion should be further regarded as a crime if it is performed without the woman's awareness or against her will, and if it is performed by someone without medical authorisation. In the contemporary discussion, the most heated debate was sparked by the paragraph according to which an abortion carried out within three months of conception was absolutely exempt from punishment.[2] The bill was again proposed in a modified form two years later. Its authors focused on clarification and explicit determination of the circumstances under which abortion could be exempt from punishment. The amendment defines five different types of indications (reasons): (1) medical, which allows an abortion if the pregnant woman's life is in danger; (2) eugenic, which at the time was understood to mean, particularly, severe nervous disorders, but also syphilis or alcoholism of one parent; (3) social; (4) if the pregnant woman was seduced and is under 16 years of age, and (5) if pregnancy was rape-related. The limit of three months after conception was preserved in the adapted version of the amendment. This text again emphasises the nature of abortion as a medical intervention, which, as such, could be performed only by a physician and in a healthcare establishment.

[2] It must be emphasised that the first proposal was not discussed in the parliament at all. The discussion of its contents took place in the general public and in the press (see e.g. Rádl 1932; Nikšová 1971, 69–82).

A third and completely new bill to annul the parts of the imperial penal code pertaining to the expulsion of a foetus was proposed in the Czechoslovak parliament in 1925. That time it was brought forward by German social democracy MPs, including Irene Kirpal, née Grundmann (1886–1977). In contrast to the previous bill, the text did not determine any temporal limit within which abortion should be exempt from punishment; however, it is more important for this text in terms of the idea that abortion should be performed not according to the will of the pregnant woman but to the opinion of a committee set up for this purpose. The committee, which should always comprise at least one woman and one doctor according to the bill, should take into consideration primarily the health and social circumstances of the pregnant woman, and, based on this assessment, they should issue a decision that would ultimately mean impunity for performing the abortion. All the aforementioned bills coincidently highlighted that one of their aims is simply to legalise a practice that was quite common and that was not regarded by people as a criminal act. However, none of these bills were passed. §§144–148 of the imperial penal code were in force until 1950. However, from the social and philosophical perspectives, the very fruitful debate also involved reflecting upon the further meaning, direction, and tasks of the women's movement (cf. Rádl 1933).

Science occupied a crucial role within the newly established republic. Similarly to most other European countries, science was expected to help revive the economy, which had been rather depleted during the First World War. In the period from 1918 to 1939, the scientific infrastructure developed dynamically and the network of scientific institutions on the Czechoslovak territory expanded. The Royal Bohemian Society of Sciences, which was already founded in the 1870s, and the Czechoslovak Academy of Sciences and Arts, which was established in 1890, both continued in their activities. Moreover, the newly founded Masaryk Academy of Labour (MAL) covered the area of technical sciences. The idea that it was necessary to establish a scientific institution that would focus on technical disciplines and produce results oriented towards and applicable in practice arose in Czech society and was already being discussed at the beginning of the 1900s, but it was not implemented until the war was over. MAL was established in 1920. A more accurate picture of the focus of MAL can be drawn based on the structure of this scientific organisation: it consisted of the Institute for Urban Construction, Institute for Emigrational and Colonisational Economic Relations, Institute for Economical Utilisation of Fuels, Institute for the Protection of Nature and the Landscape, and the Water Management Institute. At that time, technical sciences also comprised social sciences, which were developed above all in the Psychotechnical Institute where researchers dealt with issues related to the psychology of work, human productivity, and the effective organisation of labour in production (cf. Těšínská and Schwippel 2018).

Within the implementation of policies in various fields such as agriculture, industry, and foreign policy, state authorities also established specialised institutes and scientific institutions, such as the Slavonic Institute and the Oriental Institute (cf. Jůnová-Macková 2014), which indicate the direction taken by the foreign policy of the newly founded state. The scientific and specialised associations in interwar Czechoslovakia included the Czech Astronomical Society (1917), Purkyně Society for the Study of the Soul and Nerves (1919), Czechoslovak Society for Mineralogy and Geology (1923), Masaryk Sociological Society (1925), Microbiological Society in Prague (1929), and many other regional associations dealing with national history, natural science, and history. There were also informal associations conducting research; one of the best known is the Prague Linguistic Circle, founded in 1926. A specific category is represented by non-state research institutes, focusing above all on agriculture and the food processing industry (e.g. Distillery Institute, Research Institute of Sugar Refinery, Research Institute for the Use of Electricity in Agriculture). Scientific research was also carried out in private laboratories in the Baťa Shoe Company in Zlín and in the Škoda Works in Pilsen (cf. Kostlán et al. 2018). In general, the structure of scientific institutions in the 1920s and 1930s in Czechoslovakia was scattered and disorganised.

Science was to become one of the tools for spreading the good reputation of the newly established republic abroad. For this purpose, the Czechoslovak National Research Council was established in 1924. It was mainly aimed at promoting Czech science abroad, translating Czech scientific papers and books into foreign languages, and reviving international cooperation and involvement in newly established scientific structures, such as the International Research Council and the International Academic Union. After the end of the First World War, international cooperation also resumed at the level of individual contacts. In the 1920s and 1930s, a number of international congresses and conferences were held in Czechoslovakia (cf. Brádlerová 2018). In 1924, the First International Congress for the Scientific Management of Work took place in Prague. In 1927, the Third Congress of the International Union of Geodesy and Geophysics was held there. The International Thomist Conference took place in 1932. Two years later, the Eighth World Congress of Philosophy was held in Prague; its participants presented papers on logic and scienficic methodology, and papers aiming to contribute to the defence of democracy and humanism. To commemorate the 150th anniversary of the birth of Jan Evangelista Purkyně, the Fourth International Congress on the History of Science was held in 1937. With the growing scientific infrastructure and increasing research activity, the possibilities of publication also grew. New journals were published, offering more space for the publication of findings of research work and more freedom in developing various lines of thought and hence a greater variety of topics as well. An example of this is the situation in philosophy, where there was just one academic journal until the early 1920s. This journal, called *Česká mysl* [*Czech Mind*], had been published since 1900 and its articles focused on the philosophical thinking embedded in science. Its first editors, František Krejčí (1858–1934), František Drtina (1861–1925), and František Čáda (1895–1975), were positivists, yet they did not have much understanding for the then-developing epistemological and methodological issues of natural sciences. They were mistrustful of metaphysics and paid special attention mainly to psychological issues. Thus, the journal called *Ruch filosofický* [*Philosophical Action*] was founded in 1921; its founders explicitly declared in their programme their intention to fight for the freedom of Czech philosophy and against the rule of positivism. *Ruch filosofický* came out irregularly until 1942; between 1923 and 1927, it was published by the Union of Czech Mathematicians and Physicists. Until the outbreak of the Second

World War, other specialised philosophical journals were created in Czechoslovakia (e.g. *Filosofie* [*Philosophy*] and *Filosofická revue* [*Philosophical Revue*]; most were created in an effort to provide space beyond the mainstream ideological traditions and philosophical movements.

It must be emphasised that during the entire period under examination in this chapter, starting from the establishment of the Czechoslovak Republic and finishing symbolically with the closure of Czech universities in November 1939, a parallel German scientific infrastructure existed alongside the Czech one. At that time, approximately one third of the population were of German nationality. After 1918, the representatives of German universities in Prague and Brno strove to declare these universities Austrian property and students demanded that the German part of Charles-Ferdinand University be transferred from Prague to Liberec. However, none of these changes took place. The structure of the German higher education in Czechoslovakia remained unchanged, even though it had to start functioning in an absolutely different context. The German part of the university in Prague had about half the number of teachers and students that the Czech part did (cf. Bečvář et al. 1998, 181–184). In 1920, the university in Prague officially split and both institutions worked and functioned separately. In the interwar period, their communication and collaboration were not ideal, complicated by a number of disputes escalating between the two universities. One of the best-known female graduates of the German University in Prague was Gerty Theresa Cori (1896–1957), who graduated from the Faculty of Medicine; after her graduation in 1920, she left for the university in Vienna. She and her husband were awarded the Nobel Prize for their research in biochemistry in 1947.

In examining women's works in the history of Czech philosophy and science, the interwar period is characterised by women's endeavours to get involved in the existing scientific structures, gain some recognition, and participate in the education of the next generation of scholars. Hence, women's relationships with official scientific institutions appears to be the most suitable classification criterion for the given period. The following classification is not thematic but purely formal. As regards the form of women's relationship to official scientific institutions and the possibility of influencing future generations, the women can be divided into three basic groups. The first group (private female researchers in contact with scientific institutions) is composed of women conducting research more or less as a hobby. These women very often engaged in activities via which they tried to popularise the topics of their professional interest, and their links or pertinence to scientific institutions were more likely to be loose and unofficial. The second group (women at universities) includes women who managed to achieve recognition beyond their formal acceptance as members of the given scientific community and at the same time were able to present their ideas and thoughts to the next generation of scholars. This group comprises women who were authorised to lecture at universities, based on their scholarly work. The last group (solitary female thinkers) in the interwar period includes women who stood outside the structure of scientific institutions, who did not strive to be in contact with the official scientific community. These women often had no formal higher education and their works showed a strong influence of Eastern thinking and spiritualism.

4.1.2 Private Female Researchers in Contact with Scientific Institutions

Despite the increasing number of female university students and graduates in Czechoslovakia after 1918, it was very hard for women to enter the world of science and have their work accepted as something that should be shared and further developed via official educational structures. It is quite rare to find women on the lists of members of prominent learned societies in the interwar period. For example, the Royal Bohemian Society of Sciences, the oldest Czech scientific institution, admitted only two women before 1939. In 1927, Marie Skłodowska Curie (1867–1934) became a foreign member. Two years later, in 1929, Milada Paulová (1891–1970) joined the organisation as an extraordinary member. At the same time, the Czech Academy of Sciences and Arts accepted one female scientist (Marie Skłodowska Curie) and seven women writers: Eliška Krásnohorská (1847–1926), Amálie Vrbová (1863–1936), Gabriela Preissová (1862–1946), Božena Viková-Kunětická (1862–1934), Růžena Jesenská (1863–1940), and Božena Benešová (1873–1936). Albína Honzáková (1877–1973) and Alice Masaryková (1879–1966) were briefly members of MAL. It is essential for the development of science that researchers pass their ideas and the results of their scientific work on to future generations. Examples from the history of philosophy and science clearly show that the legacy can fall into oblivion despite intensive publication activities, if it is not discussed, criticised, developed, or at least commemorated within the community. The next section introduces three figures whose professional interests lay in mathematics (Marie Fabiánová), natural sciences (Marie Zdeňka Baborová-Čiháková), and the humanities (Ludmila Matiegková). These figures demonstrate the diversity of disciplines in which women in the interwar period achieved impressive results as well as the fact that the discipline in which a woman was active was not related to and did not predetermine the success that could be achieved. A much more crucial role was played by external circumstances. Each of these women had a different professional interest and different life trajectory. However, what they shared was a great gift and enthusiasm for scientific research.

Marie Fabiánová (1872–1943) was born in Železný Brod. Her father worked as a railway engineer. Very limited information is available about her life. She was one of the first girls to graduate in 1895 from the grammar school Minerva. She initially attended lectures at Prague's Charles-Ferdinand University in the Faculty of Arts as an observer. After women were allowed to enrol in degree studies at the faculty in 1897, the subjects that Fabiánová had already passed were recognised and she took the final exams in 1900, authorising her to teach mathematics and physics at secondary schools. A year later, she obtained a doctorate after defending her dissertation thesis on mathematical analysis, entitled *O rozvoji diperiodických funkcí v nekonečné součty a součiny, v řady a produkty* [*On the development of diperiodic functions in infinite sums and products, in series and sequences*]. She was a student of František Josef Studnička (1836–1903), who taught at the Faculty of Arts at Charles-Ferdinand University from 1871. Studnička was a prominent populariser of science; in the 1860s and 1870s, he gave nearly twenty lectures for women in the American Ladies' Club. In 1870, František Studnička was a founding member of the Union of Czech Mathematicians and Physicists, which is one of the oldest academic societies still in existence. The Union aimed to promote engagement in mathematics and physics and their classes at schools of all levels. At the end of the 1800s, Studnička brought Marie Fabiánová into this association; in 1903, she was the first woman to become an official member. Her place in the history of Czech mathematics and physics is connected to her article *O objevu Zeemanově* ['On Zeeman's Discovery'] (1898), which was the first article published by a woman in *Časopis pro pěstování matematiky a fysiky* [*Journal for the Pursuit of Mathematics and Physics*]. The article was based on a paper she presented at a meeting of the Union of Czech Mathematicians and Physicists when she was still a student of the Faculty of Arts. In her article, Fabiánová focused on the explanation and appraisal of the 1896 discovery by the Dutch physicist Pieter Zeeman (1865–1943). While examining the influence of strong magnetic fields on the nature of a light source, Zeeman found that the spectral lines of the source are split. In her article, Fabiánová gave a detailed and erudite description of Zeeman's experiments, highlighting that these results can be applied to solar energy. Fabiánová worked all her life as a teacher and later also as the headmaster of a secondary school. This occupation aroused her interest in the history of education. In 1909, under the pseudonym Abby Faimonová, she published a voluminous work in which she gave an overview of the Austrian and Czech school systems from the oldest times up until her present, *Dějiny školství rakouského se zvláštním zřetelem na školství české. Díl 1., Od nejstarších dob až po r. 1848; Díl 2. Po roce 1848* [*History of the Austrian School System with a Special Regard to the Czech School System. Part 1: From the Oldest Times to 1848; Part 2. After 1848*].

Marie Zdeňka Baborová-Čiháková (1877–1937) was born in Prague to the family of an primary school headmaster. Even when she was still a child, it was said that she possessed a great gift in that she could speak nine languages. While she was still a student at Minerva, a secondary school for girls, she started to attend lectures of biology, botany, and zoology at Prague's Charles-Ferdinand University. Her first academic papers were published in 1896 in the journal *Živa*. In her university studies, after she enrolled as a degree student in 1897, she focused on researching lower animals. Thanks to her expertise, Baborová was invited to participate in the work on *Otto's Encyclopaedia*, which was the most voluminous encyclopaedia of its time, published from 1888 to 1909, and co-authored by over a thousand experts. The members of its editorial board also approached women to write some entries, particularly the female students of Prague's Charles-Ferdinand University. Baborová authored the entries on Infusoria and Protozoa.

In 1901, Baborová became the first female Doctor of Philosophy; her graduation ceremony received great attention in the contemporary press, perhaps also due to her being introduced to Emperor Franz Joseph II, who was visiting Prague at that time (cf. Uhrová 2012). In her dissertation work, which was published separately in an extended and revised version under the title *Úvod k studiu o tukovém tělese členovců* [*Introduction to the Study of Corpus Adiposum in Arthropoda*] (1902), she dealt with the adipose tissue in Arthropoda. The introduction to this work states that she was inspired to investigate this topic by František Vejdovský (1849–1939), professor of zoology, embryology, and comparative anatomy at the university in Prague, who was an advocate of women's education. After graduating from university, Baborová remained in the institute directed by Vejdovský as an assistant, without remuneration. She worked there intensively until 1906, when she gave birth to her first daughter. She had known her husband, Stanislav Čihák, whom she married in 1903, since they were children. Even though he worked in financing and did not share his wife's interest in zoology, he supported her in her scientific work. Baborová was enormously influenced by her elder brother, Josef Florián Babor (1872–1951), who had a medical degree from Prague's Charles-Ferdinand University. He was interested in botany, zoology, palaeontology, and many other disciplines. From 1924, he worked as a biology professor at the university in Bratislava. Josef Babor and his sister went on many study and research stays at European scientific institutes. Together, they visited Vienna, Trieste, Wroclaw, Berlin, Munich, Paris, and Zurich. Later on, they also published together. Marie Baborová introduced the results of her research to scholarly audiences even during her university studies. A report from a meeting of the Natural Scientific Committee of the Museum of the Bohemian Kingdom states that she lectured there in 1900 on flies. In 1901, Baborová took an active part in the zoological congress held in Berlin. She authored the second part of *Velký ilustrovaný přírodopis všech tří říší* [*Great Illustrated Natural History of All Three Worlds*], edited by Jiří Janda (1865–1938) and first published in 1914, and in revised and supplemented versions also in 1931 and 1933. Baborová independently authored the parts on lower animals (worms, spiders, radiata, protozoa, etc.); she co-wrote some other chapters with her brother as well.

Of the group of women referred to as 'private researchers' in the interwar period, perhaps the most distinguishable mark was made by Ludmila Matiegková (1889–1960). Matiegková was born in Lovosice, a small town in North Bohemia, where her father, Jindřich Matiegka (1862–1941), ran a private medical practice. When Matiegková was three, her family moved to Prague. One of the main reasons for the move was her father's career and his interest in anthropology. The capital offered much better conditions for pursuing his scientific interests. In 1897, Matiegka was appointed an associate professor of anthropology and demography and started working at the university in Prague, where he later founded the Anthropological Institute. When deciding where to live, her parents also took into consideration that they could provide their daughter with a better education in Prague. Ludmila Matiegková's mother financially supported the Minerva Society, which started and ran the first grammar school for girls in the Czech lands. When Matiegková turned 12, she started studying at this school (cf. Havlůjová 2005). Ludmila Matiegková entered Minerva when its first graduates had just started to teach there, shortly after graduating from the university in Prague's Faculty of Arts (see Sect. 3.5 of this book). After passing the school leaving exam in 1908, Matiegková could not decide between studying at the Faculty of Arts or the Faculty of Medicine. Finally, she chose the first option and earned a degree in history and geography. From a very young age, Matiegková was interested in ancient Egypt; however, Egyptology was not offered as a field of study at the university when she was studying there. Nonetheless, she attended lectures on oriental philology and increased and broadened her knowledge by studying on her own. She found great support and understanding in her father. Jindřich Matiegka bought scholarly books for his daughter that were not available in the Czech lands at that time and also borrowed books for her from the university library using his registration, as they were accessible only to staff members. He also arranged short-term study stays for her at foreign universities. In 1901, Matiegková stayed in Berlin, where she attended lectures of Egyptologist Adolf Erman (1854–1937) and Assyriologist Friedrich Delitzsch (1850–1922). In the following year, she was able to attend lectures by Egyptologist Friedrich Wilhelm von Bissing (1873–1956) and Orientalist Fritz Hommel (1854–1936) at Munich University (cf. Havlůjová 2005, 71–73). In 1913, upon passing the exams authorising her to teach, Matiegková went back to the Minerva School as a teacher; she continued her teaching career until she retired. She simultaneously wrote her dissertation thesis, which was an essential prerequisite for her future scientific career.

She defended her dissertation at the Faculty of Arts of Prague's Charles-Ferdinand University in 1914 and published it two years later under the title *Názory starých Egypťanů o duši* ['Ancient Egyptian Concepts of the Soul'] (1916) in the journal of the Czech Academy of Sciences and Arts. The topic chosen by Matiegková for her work is at the intersection of philosophy, psychology, and Egyptology. Her main aim was to explore the origin and development of the notion of the soul in ancient Egyptians and compare the results with the conceptions of other cultures. Matiegková's research stemmed from the ideas of Wilhelm Wundt (1832–1920), whose conception of folk or cultural psychology (referred to as *Völkerpsychologie*) was widely discussed at that time. Matiegková's work 'Ancient Egyptian Concepts of the Soul' is founded in a well-considered methodology and shows the author's admirable erudition. Matiegková points out the drawbacks of the approach of older researchers, who referred in their research of ancient Egyptian concepts to the accounts of ancient philosophers and writers. In contrast, Matiegková's works refer to original sources, originating from the time when Egyptians were not in contact with other cultures. Moreover, as her aim was to explore popular concepts of the soul, she concentrated on the analysis of written and archaeological materials and religious and funeral rites. In her opinion, the popular concept of the soul must be sought not only in poetry and historical and medical works but also in popular myths, sayings, and practices. Matiegková regarded the comparative perspective offered by Wundt in his folk psychology as the necessary starting point of her research; at the same time, she was aware of its drawbacks. She believed that the comparative perspective is of greater value than mere speculation, although it can also produce inaccurate results, as it is based on analogy method. Ancient Egypt was the lifelong focus of Matiegková's professional interest. In the 1920s, she went on several private research trips to Egypt and Palestine. She presented her findings and experiences in popular scientific works, fiction, and travelogues, including *Jak vzniklo písmo* [*On the Origin of Writing*] (1918), *O starověkých bozích a hrdinech* [*On Ancient Gods and Heroes*] (1925), *Záhada* [*Mystery*] (1934), *V objetí sfingy* [*In the Arms of the Sphinx*] (1927), and *Dítě v starém Egyptě* [*Child in Ancient Egypt*] (1937).

In the period after she obtained her doctorate, Matiegková's work was linked to that of her father, who, in addition to historic and prehistoric anthropology, engaged in anthropological research of the population in the Czech lands. Matiegková helped her father examine the remains of renowned figures such as Jan Amos Comenius and Tycho Brahe. In the 1920s and 1930s, she published several independent studies in the journal *Anthropology*, including *Vyšetřování egyptských mumií* ['Examining Egyptian Mummies'] (1929); *Tělesná zdatnost starých Egypťanů* ['Physical Fitness of Ancient Egyptians'] (1933), and *Rozlišování plemen a jeho praktické důsledky v starém Egyptě* ['Distinguishing Races and Its Practical Consequences in Ancient Egypt'] (1935). Matiegková was also close to the Egyptological Seminar, which was founded in 1925 at the university in Prague by František Lexa (1876–1960), one of the most prominent figures of Czech interwar Egyptology. Lexa proposed Matiegková as a member of the Oriental Institute, which she joined in 1930 (Havlůjová 2005, 121). Matiegková retired from her career as a private researcher in the 1950s, after she published studies written in German in the journal *Oriental Archive*, in which she dealt with the history of Egyptian medicine and pharmacology (*Drei organische Adstringentien in den altägyptischen medizinischen Papyri* ['Three Organic Astringents in Ancient Egyptian Medical Papyri'] (1952), *Tierbestandleile in den altägyptischen Arzneien* ['Animal Parts in Ancient Egyptian Medicines'] (1958), and *Produkce tieriches Exkretion un Sekretion in den altägyptischen Heilmitteln* ['Animal Excretion and Secretion in Ancient Egyptian Remedies'] (1959).

4.1.3 Women at Universities

In the interwar period, women worked at all of the faculties of Charles University except for the Faculty of Theology. This chapter focuses on the women who were recognised by the scientific community and authorised to conduct university teaching (the venia legendi). Most associate professors (docents) could be found at the Faculty of Arts. In 1925, Milada Paulová (1891–1970) became an associate professor of general history of Eastern Europe and the Balkan Peninsula. Four years later, Flora Kleischnitzová (1891–1946) earned an associate professorship for her work in the history of Slovak literature. In 1930, Růžena Vacková (1901–1982) was awarded an associate professorship in classical archaeology, and Drahomíra Stránská (1899–1964) was authorised to give lectures in ethnography in 1932. At the Faculty of Science, which was founded in 1920 after it split from the Faculty of Arts, two women became associate professors in the 1930s: Albína Dratvová in 1932 in the philosophy of natural sciences and Julie Moschelesová in 1937 in anthropogeography. In the interwar period, two associate professors were also at the Medical Faculty: Vlasta Říhová-Knappová (1890–1960), who obtained her associate professorship in dermato-venerology in 1932, and Olga Valentová-Denigerová (1900–1981), who obtained hers in 1933. Just one woman was authorised to lecture at the Faculty of Law, Jarmila Veselá (1899–1972), who became an associate professor of penal law in 1928 and then worked in the Institute of Criminology, which was the first and for a long time also the only scientific institute of the Faculty of Law. The following text presents a more detailed look at the lives and works of three of these nine associate professors. Milada Paulová was the most successful from the perspective of her further career. In 1935, she was appointed an adjunct professor, which was not accomplished by any other woman before the Second World War. Julie Moschelesová and Albína Dratvová aptly illustrate how complicated the relations between the Czech and German scientific communities were.

Milada Paulová (1891–1970) was born in Loukov and graduated from a school for girls preparing future teachers in Prague. She and her friend and classmate Albína Dratvová passed the school leaving exam as external students at the Academic Grammar School in Prague and enrolled in Prague's Charles-Ferdinand University in 1913. Paulová studied history and geography and became a student of Professor Bidlo (cf. Havlíková 2009). In 1918, she earned a doctoral degree based on defending her work *Styky českých husitů s cařihradskou církví na základě poměrů byzantských* [*Encounters of Czech Hussites with the Church in Constantinople on the Grounds of Byzantine Conditions*]. She was employed in the Public and University Library until 1935; at the same time, she worked on her scholarly texts and went on study trips.

Paulová became the first female associate professor (1925) and adjunct professor (1935) in the Czech lands. After the Second World War, she obtained full professorship. She focused on the history of Southeast Europe and Byzantology and worked as a professor at the Faculty of Arts of Charles University until 1961. She was also a devoted editor; her work for the prestigious journal *Byzantinoslavica* is of great importance. It was especially to her credit that the journal continued even after the Second World War. Paulová went on a study stay to Paris in 1927 and was in touch with foreign Byzantologists from that time. In her texts on Byzantology, she dealt e.g. with the issue of Czech-Byzantine relations under the rule of the Bohemian King Ottokar II and the influence of Islam on the territory conquered by Turks—her work *L'Islam et la civilisation méditerranéenne*, written in French and published by the Bulletin of the Royal Bohemian Society of Sciences in 1933, critically develops the conception of the Belgian historian H. Pirenne and deals with the role of Islam in the development of European civilisation.

In terms of modern history, Paulová is well known for her works on the origin and backdrop of the foundation of the new Czechoslovak state, on the relations between Czechoslovakia and Yugoslavia, and on Czech and Yugoslav resistance in the First World War. She collected material during her study stay in Yugoslavia in 1920–1921. Her work dealing with this issue was published in Croatian in 1925 as *Jugoslavenski odbor: Povijest jugoslavenske emigracije za svjetskog rata od 1914–1918* and Paulová submitted it at university as her habilitation work. Paulová became an internationally recognised expert in Yugoslav history. She authored the works *Jihoslovanský odboj a česká Maffie I* [*Yugoslav Resistance and Czech Maffia I*] (1928), *Tomáš G. Masaryk a Jihoslované* [*Tomas G. Masaryk and Yugoslavs*] (1938), and *Dějiny Maffie: Odboj Čechů a Jihoslovanů za světové války 1914–1918 I, II* [*History of Maffia: Czech and Yugoslav Resistance in the World War, 1914–1918 I, II*] (1937–1939).

In 1929, Paulová became an extraordinary member of the Royal Bohemian Society of Sciences and in 1946 an extraordinary member of the Czech Academy of Sciences and Arts: she was one of the first women in Czechoslovakia to join these institutions. At the end of her life, Paulová returned to the Yugoslav issue and revised and published her monograph *Tajný výbor (Maffie) a spolupráce s Jihoslovany v letech 1916–1918* [*Secret Committee of Maffia and Cooperation with Yugoslavs 1916–1918*] (1968).

Julie Moschelesová (1892–1956) was born in Prague; however, as her mother was blind, she was raised and educated in London, with the family of her uncle, Felix Stone Moscheles. Her further development and professional interest were profoundly shaped by her travels all over Europe and North Africa, during which she met the Norwegian geologist Hans Henrik Reusch (1852–1922). Thanks to his help, she was employed as a secretary and translator in the geological institute in Oslo where she started to take a keen interest in geomorphology. Another important encounter took place in 1912, when Moschelesová met Alfred Grund (1875–1914), a professor of geography at the German Prague's Charles-Ferdinand University who was on a study stay in Norway. Grund noticed her gift and invited her to return to Prague and start studying at the university under his supervision. Moschelesová enrolled in the Faculty of Arts of the German University in Prague in autumn of that year, attending lectures of geography, geology, and meteorology. She completed her studies in 1916 when she passed the state exam and defended her dissertation work *Die postglazialzeit in Skandinawien* [*The Postglacial Period in Scandinavia*], in which she summarised the findings of her geomorphological research of the Scandinavian peninsula, which she had already started during her time in Oslo.

Even though she managed to obtain an assistant position at the German University, she left in 1922 due to nationalist and racists attacks and went to the Geographical Institute of the Czech Charles University, where she was appointed an associate professor in 1934. Because she was Jewish, she fled the country before the 1939 occupation of Czechoslovakia by the German army. Thanks to the support of her friends she managed to escape to Australia, where she was able to teach at the University in Melbourne. An interesting and not very researched episode in Moschelesová's biography is her work in the geographical department of the army of the Dutch East Indies, which she joined after this colony had been attacked by Japan (cf. Martínek 2010, 185). Shortly after the end of the Second World War, in 1946, Moschelesová returned to Czechoslovakia; however, she did not find any of her relatives there, as her family members had been murdered by Nazi's, nor did she find good conditions for life and scholarly work. She again became an assistant at the Faculty of Science of Charles University, but she had a very low salary. To make ends meet, she worked a lot as a translator. Her material situation improved in 1953 when she obtained state associate professorship. By that time, she was suffering many serious health problems, and she died three years later.

Julie Moschelesová's extensive scholarly works present two main lines of interest. Moschelesová dealt with physical geography and the relationship between geography and people's social and economic activities (anthropogeography or social geography). Works in which she explored the condition and principles of the development of individual parts of a landscape fall within the period of her work at the German university in Prague. Moschelesová examined the climate in Prague and on the territory of the Balkan Peninsula (*Das Klima von Prag*, 1917; *Das Klima: von Bosnien und der Hercegovina*, 1918) and, above all, the geomorphology of the Czech territory in *Morfologické důkazy nejmladších tektonických pohybů v severozápadních Čechách* [*Morphological Proofs of the Youngest Crustal Movements in Northwest Bohemia*] (1921) and *On Young Crustal Movements in the Carpathians* (1923). She stressed the influence of the geological structure and crustal movements. Her hypotheses on the courses of some Czech and Moravian rivers in various periods and dating of some Czech volcanic rocks are still valid today (cf. Martínek 2010, 181). After she had left the German university for the Czech university, she started to pay more attention to the newly established discipline of social geography, in which she obtained her associate professorship in the 1930s. In this area, she is recognised mainly for her works *Wirtschaftsgeographie der Tschechoslowakischen Republik* (1921) and *Landeskunde der Britischen Inseln* (1925). She was also interested in the theoretical issues of the discipline, as shown by her article *Logická soustava zeměpisu člověka* ['Logical System of Man's Geography'] (1925) and in selected topics from the regional geography of the world, as seen in her article *O Darwinově teorii vzniku korálových útesů* ['On Darwin's Thesis of the Origin of Coral Reefs'] (1921). Moschelesová published a major part of her findings abroad where she was also a respected researcher. She was an honorary member of the Sociological Society in London. The French organisation Société de Géographie commerciale de Paris awarded her a gold medal in 1930.

Albína Dratvová (1892–1969) was born in Prague; her father worked as a court clerk. In 1911, she graduated from the state institute for female teachers; however, as she wanted to continue studying at university, she started to privately prepare for her school leaving exam. She passed this exam at the Academic Grammar School for Boys and enrolled in the Faculty of Arts at the Czech Prague Charles-Ferdinand University in 1913 where she studied mathematics and physics. Upon completing her studies in 1918 by defending a dissertation work in which she dealt with Descartes' ethics, Dratvová taught at grammar schools for girls in Prague. According to the diary she kept between 1921 and 1961, she was offered a position at the Psychotechnical Institute in the Masaryk Academy of Labour in early 1924. Dratvová had been trying to obtain such a position in which she could be fully engaged in scientific research, and she had prepared for it by studying statistical methods and calculations of probabilities. However, for some reason she eventually turned down the offer to work in the Psychotechnical Institute. Several notes in her diary entries from that period express her doubts about using experimental methods and measures when studying psychological processes and about the reasonableness of treating the results acquired in such a way just as one treats random phenomena (Dratvová 2008, 66–76). Dratvová's career as a secondary school teacher ended in 1928 when she obtained the post of a state clerk and was hired in the pedagogical department of the Ministry of Education and National Enlightenment where she worked until she retired early in 1942.

In the interwar period, the focus of Albína Dratvová's professional interest was the philosophy of science.[3] Dratvová concentrated on the criticism of positivism, the prevailing approach in the first half of the 1900s towards determining the relationship between philosophy and science and towards solving the philosophical questions of science. She formulated the main theses of her ideas in her article *Positivismus ve fysice* ['Positivism in Physics'], which came out in 1924 in the journal *Ruch filosofický*; it was also published separately in the same year. Dratvová tried to prove that positivists fail to implement the programme they themselves set up, based on the principle of working only with immediately perceptible facts, repudiating all that is metaphysical. In her critical analysis, she focused mainly on the philosophy of Ernst Mach (1838–1916), since she was interested in finding how in particular the positivist approach is reflected in physics. Dratvová believed that it is not possible to eliminate metaphysical hypotheses when examining the world and that some of the hypotheses (e.g. the precondition of the isomorphism of natural phenomena and the precondition of the existence of a world independent of the observer) are tacitly presupposed within positivism and play a major role in the development of science. Dratvová disapproved of the idea that there are facts immediately given through observation and that the aim of science should be to look for permanent dependencies between these facts, i.e. to formulate universal laws via the tools of mathematics. According to Dratvová, the main aim of science is to explain the phenomena around us, not only describe and categorise them.

The development of modern physics, in particular quantum mechanics, raised *inter alia* other doubts pertaining to the role of causality in science. Dratvová dealt with this issue in depth in her work *Problém kausality ve fyzice* [*The Problem of Causality in Physics*] (1931), with which she habilitated a year later. Hence, in 1932 she became the first female associate professor at the Charles University Faculty of Science.[4] The associate professor position came with the opportunity to deliver independent lectures at the university. Albína Dratvová lectured at the Faculty of Science until the closure of Czech universities in autumn 1939. *Problém kausality ve fyzice* offers a comparison of the classical and modern conceptions of causality. Dratvová shows that while classical physics is characterised by an absolute perception of causality in which the relationship between the cause and the effect is defined as clear-cut, the world itself appears to be indefinite and random at the microscopical level. Dratvová claims that behind the classical conception of causality, a presumption of the permanency of a substance is given by the fact that the elementary particles constituting the world can be delimited only by its speed and position. This presumption was disproved by quantum mechanics; the determinist description was replaced by statistical laws and the probability theory. However, Dratvová argues it does not mean causality ceased to be valid as a pragmatic and useful postulate of scientific research. Dratvová regarded

[3]Dratvová's works published during and after the Second World War will be introduced in the following subchapters.

[4]In 1920, a law was adopted determining the Czech part of the original Charles University to be its successor. After that, the German section used the name German University in Prague. This brought the division of the university, initiated in 1882, to an end.

causality as one of the irreducible preconditions of physical research, which has both a heuristic and an epistemological role. It was the heuristic function of these preconditions among which she included, besides causality, the preconditions of simplicity, finality, and sensory illustrativeness of natural phenomena, which she addressed in her work *Heuristické předpoklady fysikálního bádání* [*Heuristic Preconditions of Physical Research*] (1934). It is evident from her diary that the ideas she formulated in this work became the basis of a paper she was planning to deliver at the international philosophical congress in 1934 in Prague. However, as the number of Czech active participants was limited, Dratvová ultimately did not deliver her presentation. In the second half of the 1930s, she published the works *O aplikabilitě matematiky* [*On the Applicability of Mathematics*] (1936) and *Planckova filosofie* [*Planck's Philosophy*] (1939), in which she introduced the philosophical ideas of Max Planck (1858–1947), which resembled her ideas in particular due to their realism and approach to the issue of causality. Albína Dratvová's comprehensive work is surely *Filosofie a přírodovědecké poznání* [*Philosophy and Scientific Knowledge*] (1939), in which she provides systematic insight into the logical, methodological, epistemological, and metaphysical problems of natural sciences. Dratvová criticised positivism in its various forms; she criticised Mach's phenomenalism and Carnap's physicalism. She viewed scientific research as a never-ending activity. As one of her main arguments against positivism was related to the development of science, it is no surprise that Dratvová also paid attention to the history of science. In several parts of her works, she pointed out that the philosophy of science underestimates lessons learned in the past. At the international conference on the history of science held in 1937 in Prague, she presented her paper 'Newton as a Philosopher', in which she focused on the role of god in Newton's philosophical and scientific system. In 1942, she published two journal studies on key figures from the history of science and philosophy, to commemorate 300 years since Galileo's death and at the same time 300 years since the birth of Isaac Newton. Both studies were published in the journal *Česká mysl*.

Before Dratvová focused on the philosophical and methodological issues of natural sciences, in the first half of the 1920s, she concentrated on psychological topics. In this period, she published two works dealing with the question of subconsciousness and a number of papers and reviews of books about psychology and psychological phenomena. The work *O stavech podvědomých a Freudově psychoanalyse* [*On Subconscious States and Freud's Psychoanalysis*], (1921) is a transcript of the lecture Dratvová delivered in the Philosophical Union in a cycle about the human soul. For her lecture, Dratvová chose the topic of subconsciousness, which was a frequently discussed and not universally accepted concept at that time. One of the opponents of subconsciousness was Mihajlo Rostohar (1878–1966), an experimental psychologist from Slovenia who studied under the supervision of Alexius Meinong at the university in Graz and spent some time studying at Wilhelm Wundt's psychological institute in Leipzig. Rostohar lectured at the university in Prague from 1910; later he started to teach at the newly established university in Brno. His negative attitude to the concept of subconsciousness and Freud's psychoanalysis were made clear in the debate that followed Albína Dratvová's lecture. Dratvová later developed the given topic more profoundly in her work *O tajích podvědomí* [*On the Mysteries of Subconsciousness*] (1925). Although she did not deal with psychological topics later in her academic career, she kept up with the development in this discipline, as she regarded it an important part of the theory of cognition. This attitude shows the influence of her teacher František Čáda (1865–1918), who was engaged in pedagogy and psychology and, as regards philosophical disciplines, predominantly in the theory of cognition and ethics. Čáda was an advocate and supporter of women's studies at universities. He also dealt with the topic in his works. In 1911, he published an article on female representatives in ancient philosophy (cf. Čáda 1911).

The last field mentioned in connection to Albína Dratvová's interwar activities pertains to the didactics of philosophy, a subject she came to through her teaching career and her subsequent work in the Ministry of Education. Dratvová authored several textbooks on philosophical propaedeutic, including logic and psychology, *Úvod do filosofie* [*Introduction to Philosophy*] (1928) and *Filosofie* [*Philosophy*] (1936). Her textbooks were very popular and translated into Slovak, Hungarian, and French. After the end of the Second World War and the communist party takeover, they were publicly rejected, as they did not correspond to the principles of Marxism-Leninism. Albína Dratvová was a respected philosopher of the interwar era. Her work was formally awarded in 1946, when she became an extraordinary member of the Royal Bohemian Society of Sciences.

4.1.4 Solitary Female Thinkers

During the first Czechoslovak Republic, two female thinkers, Pavla Moudrá and Anna Pammrová, stood outside the official scientific or academic institutions and structures but continued with their publication activities. Both women were from the generation that had had no access to secondary or higher education (the first grammar school for girls was founded in 1890; women were not allowed to enrol in degree studies at the faculties of arts until 1897). In their works, there is an interplay of various intellectual influences, which both women learned primarily by themselves. Moudrá and Pammrová were friends; they kept in touch through letters. Their views and attitudes have some common points of departures and intersections; however, their outcomes and results are different. Both thinkers criticised the clerical form of religion and its dogmas; they were influenced by occultism, spiritualism, theosophy, and ancient Eastern teachings. Nevertheless, they varied in their evaluation of culture, the future development of society, and women's role in this process.

Eastern thinking started to penetrate the Czech environment approximately in the 1870s, mainly thanks to František Čupr (1821–1882), who introduced the Czech reader to the basic texts of Indian and Chinese philosophy in his four-volume work *Učení staroindické* [*Teachings of Ancient India*], published between 1876 and 1881. Moreover, he translated the *Bhagavad Gita* and *Tao Te Ching*. Čupr's work stemmed from the principles of Herbartian philosophy, and his interest in Eastern teachings was sparked by the works of Arthur Schopenhauer. Čupr was interested in the ancient Indian teachings particularly from the perspective of their importance for the origin and development of Christian concepts; he thus viewed it within the philosophy of religion. In contrast, Moudrá and Pammrová were attracted by Eastern thinking predominantly for its simultaneously profound and unrestrained spirituality, which, as they believed, could give meaning even to seemingly pointless historical processes. At that time, the synthesis of Eastern and Christian religions, represented by the theosophical movement of the Russian countess Helena Petrovna Blavatsky (1831–1891), enjoyed particular popularity in the Czech lands. The growing interest in various forms of spiritualism and occultism that occurred in the late 1800s and early 1900s went hand in hand with the end of the Industrial Revolution, the rapid development of science and other processes of modernisation and societal secularisation that disrupted the traditional sources of certainty, and the legitimisation of the social and world order. The events of the First World War intensified feelings of crisis; thus, interest in spiritualism and occultism peaked during the First Republic.

Pavla Moudrá (1861–1940) was born and died in Prague; however, she spent a major part of her productive life outside the centre of the action, in Neveklov, a small town about 50 km to the south of Prague. She received her education at a secondary school for girls and at private institutions focused on teaching foreign languages. Later on, Moudrá made excellent use of her language skills in her many translations. In 1885, she married writer and journalist Pavel Albieri, born Jan Mucek (1861–1901), whom she assisted in editing the regional weekly newspaper *Ratibor*. Their marriage, into which their only son was born (he died as an infant), ended in divorce four years later. In 1898, Pavla Moudrá tried to publish her own women's magazine with no success. After this enterprise failed, she left for Russia where she briefly worked as a governess. In the 1890s, upon returning to Prague, Moudrá was very active in the Prague society for animal protection and wrote short texts, reviews, and news from the world of theatre for women's magazines under her pen name Olga Přibylová. In 1902, she got married for the second time, to Antonín Karel Mrha (1875–1947), with whom she moved to Neveklov where they ran a convenience shop. However, her second marriage did not work out well either. Even though they were not officially divorced until the late 1920s, the couple separated by the end of the First World War (cf. Pánek 1974). Pavla Moudrá engaged in translating, writing, and public activities. She co-launched the Czech peace movement with Jindřiška Wurmová (1864–1953). In 1902, Moudrá joined the Theosophical Society in Prague. She advocated the principles of theosophy in various discussions (see e.g., the discussions on the legitimacy of using animals for scientific purposes in Sect. 3.5). The wide range of her interests is aptly illustrated in the collections from her lectures for the public, published in 1918 and 1919, *Výbor přednášek. Díl 1* [*Selected Lectures. Part 1*] and *Výbor přednášek. Díl 2* [*Selected Lectures. Part 2*]. In them, Moudrá spoke out against war, alcoholism, and prostitution, and she formulated her concept of the new man and new culture.

Today, Pavla Moudrá is known as a Czech journalist, translator, and children's author. Her translating activities were very much motivated by financial reasons. Moudrá translated from English, French, Russian, and Swedish. In terms of the sheer number of translations, her favourite and most translated author was Rudyard Kipling.

Nevertheless, to be able to understand her ideological profile, her translations of non-fiction must be mentioned as well. Moudrá translated *The Difficulties of Individualism* by Sidney James Webb (1859–1947), *Historical Letters* by Pyotr Lavrov (1823–1900), and biographies of Mahatma Gandhi (1869–1948) and Annie Besant (1847–1933). In terms of Czech thinkers, Pavla Moudrá was most influenced by one of the best-known representatives of the Bohemian reformation, Petr Chelčický (c. 1390–c. 1460), and by philosopher and pedagogue Jan Amos Comenius (1592–1670). The focus of Pavla Moudrá's works for children and young people stems from her belief that education plays the most crucial role in the moral renewal of society. Moudrá developed her concept of the new man who will be the bearer of a new culture within theosophy and faith in reincarnation and revival. The aim of this revival process, to which everything that is alive is subordinated, should be to accomplish self-knowledge, enlightenment, and a merging with god. Moudrá was convinced that women had an irreplaceable role in this process as mothers and educators. She published her ideas in her works: *O potřebě reformy výchovy* [*On the Need to Reform Education*] (1920), *Poslání ženy ve světle theosofie* [*Woman's Mission in the Light of Theosophy*] (1922), *Můj odkaz světu* [*My Legacy to the World*] (1925), *Obrození duší* [*Revival of Souls*] (1926), and *Za světlem Komenského* [*Towards Comenius's Light*] (1933). Pavla Moudrá's works can be characterised as typical eclecticism and syncretism, combining many varied and often contradictory intellectual influences. In comparison to Anna Pammrová, whose work is much more original and offers a more critical view of society and the contemporary feminist movement, Moudrá views the future with optimism. This optimism results in her intensive effort to actively participate in the public space, join various discussions and disputes, and mobilise the public for the upcoming revival.

Anna Pammrová (1860–1945) was born as the oldest of four siblings in a forester's family. She spent her childhood and youth in a foster family, where she was sent when she was nearly six years old, probably due to her parents' bad financial situation. In her autobiography, which she called *Antieva*, she described this as a very harsh period. Although her education was probably haphazard, Pammrová had a very good knowledge of foreign languages. In the early 1880s, she left for Vienna to work as a private governess. After several years, she returned to her parents in Moravia, where she coped with the challenges of being a single mother. She refused to marry the father of her daughter, Leopold Stöger (cf. Kremenová 2005). After some time, she gave into familial pressure and in 1893 she married František Kroh, who was a forester, just like her father. The marriage, during which she gave birth to her second child, officially ended in divorce in 1903. Already in 1899, Anna Pammrová had sought refuge in a forest seclusion, where she lived in very modest conditions until her death. An exception to her very rare encounters with relatives and friends was her contact with the Havel family (including Václav Havel, who later became president), who had owned a summer residence close to where she lived since 1907.

Anna Pammrová was influenced by Ancient Indian wisdom, theosophy, occultism, and the philosophy of Rousseau, Tolstoy, Schopenhauer, and Nietzsche. Moreover, she was decisively influenced by Theodor Lessing (1872–1933), some of whose works she translated into Czech, including his *Europa und Asien* [*Europe and Asia*]

(1918), *Geschichte als Sinngebung des Sinnlosen* [*History as Giving Meaning to the Meaningless*] (1919), and *Die verfluchte Kultur* [*Cursed Culture*] (1922). Pammrová was inspired above all by Lessing's radical criticism of modern European culture and civilisation, focused particularly on criticising its rationality and colonisation tendencies. In terms of Slovak and Czech figures, she was most influenced by evangelical preacher and religious thinker Ján Maliarik (1869–1946) and the poet and renowned Czech symbolist Otokar Březina (1868–1929). Pammrová's correspondence with Březina started after they met in 1887 and lasted for over 40 years. They met in person only three times. Březina's letters to Pammrová started to be published almost immediately after the poet's death in the 1930s.

Anna Pammrová's works, notably *Alfa. Embryonální pokus o řešení ženské otázky* [*Alpha. Embryonic Attempt to Solve the Women's Issue*] (1917), *Mateřství a pamateřství* [*On Motherhood and Pseudo-motherhood*] (1919), *Cestou k zářnému cíli* [*On the Way to the Unique Aim*] 1925], and *Zápisky nečitelné* [*Illegible Notes*] (1936), present three main themes. These themes, which are mutually interrelated and complementary, are: criticism of modern society, woman's role and women's emancipation, and an ecological appeal for the reform of anti-nature thinking (cf. Gabriel et al. 1998, 435). Pammrová was convinced that the idea of progress, the basis of European society and science, was mistaken and that one of the main sources of the crisis of culture and civilisation, manifesting itself, *inter alia*, in the prevailing consumerism and social injustice, was the rule of man. However, she also assigns major responsibility in this historical process to woman, who, in Pammrová's opinion, betrayed her original mission as the protector and caretaker of life. She believes that women have contributed to their own enslavement and to the overall crisis of humanity and civilisation by submitting to flesh and sexuality. She sees the suffering and miseries caused by constant reproduction, accompanied by growing pressure on nature, as pointless and redundant, unable to be a source of redemption or liberation. She argues that the only solution is a radical rejection of civilisation and the establishment of a new, more intense relationship with nature, based on spiritual principles. Pammrová did not draw a very clear picture of her vision of this 'unique aim' or provide clear instructions on how to reach it; her ideas nevertheless became an inspiration for the current feminist ecosophy (cf. Jemelka 2016, 98–104).

4.2 Second World War

Events that started with the Munich Agreement in September 1938 and continued through the creation of an independent Slovak state and the occupation of the rest of the Czech lands on 15 March 1939 put an end to democratic Czechoslovakia and meant the instalment of the occupation regime in Bohemia and Moravia. Czech science and philosophy operated in the extraordinary situation of a nation whose very existence was under threat.[5] Political and cultural life in the protectorate[6] was supervised and controlled. The resolution of the protectorate government on 6 April 1939 proclaimed *Národní souručenství* [National Partnership] the only political association of the Czech nation within the territory of the protectorate. The National Partnership unified members of various pre-Munich political parties, yet it was not and was not allowed to be a real political party. Its programme was educating the nation to be unified in the spirit of Christianity and nationalism as opposed to liberalism and communism. The Cultural Council, originally established in December 1938, became the body of the National Partnership in March 1939. Its chair was a literary historian, university professor Miloslav Hýsek (1885–1957). The Cultural Council resolved to assist in the development of a Czech culture that would be apolitical, void of any ideology and any influence of philosophy. The national culture, deprived of undesirable philosophical influences, was to be developed under the auspices of the Greater German Reich, which was pronounced the guarantor of Czech existence. The persecution of the Jewish people started after the Munich Agreement, and there were many victims in the Jewish communities in Czech lands. Some scholars of Jewish origin managed to emigrate (e.g. geographer Julia Moschelesová[7]), but many Jewish intellectuals were murdered by the Nazi regime (cf. Rothkirchen 2006).

The conditions in the protectorate changed on 27 September 1941, when Reinhard Heydrich took up the post of Deputy Reich Protector. The Prime Minister of the protectorate government at that time, General Alois Eliáš, was arrested and sentenced to death for his cooperation with domestic resistance and for contacts with the Czechoslovak representation in London. Immediately after that, on 28 September, martial law was introduced and the first wave of Heydrich terror started, with mass arrests and executions of Czech citizens: soldiers, teachers, clerks, politicians and party officials, especially communists, white-collar workers, members of *Sokol Organisation*, etc. In December 1941, paratroopers were airlifted in from England. On 27 May 1942, Jan Kubiš and Jozef Gabčík assassinated Heydrich, who died on 4 June. The introduction of martial law launched the second wave of Heydrich terror in the protectorate; 3,188 people had been executed by 30 July. Heydrich's assassins shot themselves on 18 June after a fight in the Orthodox Church of Saints Cyril and Methodius.

The fact that the Czech lands suffered from oppression and a lack of freedom during the Protectorate of Bohemia and Moravia did not mean the complete elimination of Czech intellectuals' activity. Most of those who remained in the region refused to let others silence them, even though their chances of conducting research and publishing were much slimmer. The period of the occupation of the Czech lands can be viewed as an interim period after the cultural flourish of the 1920s and 1930s and before the transformation of power of the late 1940s. In the Czech thinking during the occupation, there was a lack of trust in German speculative philosophy, alongside

[5] On Czech philosophy in the protectorate period, cf. Zouhar et al. (2006, 2007).

[6] *The Protectorate of Bohemia and Moravia* was the official name given by the German administration.

[7] For the overview of Moschelesová´s life and work see Sect. 4.1.3 Women at Universities.

an increased art of compromise, hidden meanings, allegories, and openness to leftist orientation. The experience of the war and the occupation significantly affected the intellectual and social activities of a number of figures of philosophical life in the post-war period and undoubtedly shaped the spiritual and political dimensions of the later national life.

The Czech philosophy always considered the issues of the Czech nation's intellectual and cultural orientation and the problems of the Czech society's relation to the historical transnational context. It was also expected to respond to contemporary social problems in the complicated situation of the German occupation of the Czech lands and the war years. Hence the Czech philosophical thinking maintained the traditional thematic subjects even in the protectorate. In his article *Filosofie v dnešní situaci* ['Philosophy in Today's Situation' (1939), Jan Patočka rejected the idea that the historical role of philosophy was already over; at a time of philosophical decline, he appealed for philosophical education, arguing that all the spiritual wealth up to then should become part of the Czech national heritage and, in the given situation, also of national restoration. The history of European and Czech philosophy was paid great attention in the protectorate period; in particular, there were studies dedicated to the works of J. A. Comenius and translations of the major works of the European philosophical tradition.

Nevertheless, during the war period, the space in which Czech science and philosophy functioned changed enormously and narrowed considerably. Before the war, the public influence of philosophy occurred not only at universities, but also at secondary schools, in journals and cultural magazines and in the daily press, in original book productions and in translated literature. The protectorate years of 1939–1945 meant severe restrictions in this respect. In November 1939, Czech universities were closed down and so were the philosophical seminars of the Faculties of Arts of Charles University in Prague and of Masaryk University in Brno, seminars for methodology and the history of natural and exact sciences at the Faculty of Science in Prague and other institutes of related disciplines (sociology, psychology, aesthetics, pedagogy, etc.) at the aforementioned universities. Up until the end of the war, only the Prague's German University (cf. Míšková 2002)[8] and German technical universities in Prague and Brno remained open. Some university teachers went into early retirement, younger teachers started to teach at grammar schools (Patočka, Popelová) or were employed in the system of state authorities (Dratvová). In some cases, experts from technical colleges could find work in specialised institutes. For example, after the Czech universities were closed down, the chemist Julie Hamáčková started to work for the National Institute for Public Health in Prague, where she founded a laboratory of hydrochemistry and examined the issue of river pollution; the radiologist Adéla Kochanovská-Němejcová worked in a research laboratory in the Škoda Works in Pilsen.

[8] At the German university in Prague, lectures on philosophy were given in 1939–1945 by Theodor Czermak, Hans Richard Bernard Günther, Walter Heinrich, Kurt Schilling, Anton Stonner, Hans Julius Ernst Wolff, and Josef Zaus.

Many Czech scholars went into exile or participated in the resistance against Germans. The Czech philosophers spending the war in exile included Jan Blahoslav Kozák, who lived in the USA, and J. L. Fischer, who spent the war illegally in the Netherlands. One representative of the resistance was the classical philologist and art theoretician Růžena Vacková, who was accused of treason in February 1945; the end of war saved her from execution. Another well-known scholar participating in the resistance was the philosopher Božena Komárková, who was involved in the resistance organisation *Obrana národa* [Defence of the Nation] at the beginning of the occupation (cf. Havelková 2012). She was arrested in early 1942; after two years in custody, she was sentenced to 12 years for treason. She was released from prison in February 1945 by the Red Army. Thanks to prison chaplains, Božena Komárková had books at her disposal in prison; thus she was able to study Plato's and Augustine's works. She then used this study in her dissertation work *Obec Platónova a Augustinova* [*The City of Plato and Augustine*] at the Faculty of Arts in Brno in 1947.

The restriction of Czech philosophical life was evident also in philosophical journals. The most important pre-war Czech philosophical journal *Česká mysl* was published only in 1941–1944 (editors and authors of war contributions included Albína Dratvová and Jiřina Popelová); the *Ruch filosofický*, which was founded later, came out only in 1939 and 1941/42. Despite that, the war issues of both journals comprised some important contributions to Czech philosophy, even though critical discussion was largely limited to the review sections of both journals. The activity of the pre-war organisation *Jednota filosofická* [Philosophical Union] was interrupted after March 1939 to protest against the German occupation.[9] All publications were subject to German supervision and interventions. The general lack of journals and the limited scope of those that were being published led publishers to bring out editions comprising texts of the size of more extensive studies. An example of this was an edition by the publisher Václav Petr called *Svazky úvah a studií* [*Volumes of Ideas and Studies*], the first volume of which came out after 15 March 1939.

During the protectorate period, female scholars found themselves in a particularly difficult situation, as they were disadvantaged not only in terms of their gender, but also as members of a nation whose cultural and academic life was to be gradually eliminated. Despite that, two significant representatives of pre-war philosophy, Albína Dratvová and Jiřina Popelová, carried on with their studies and publication activities, even in these harsh conditions.

After leaving university, Jiřina Popelová taught at a grammar school in Prague; thus, she stayed in touch with students. Her former students remembered her open approach and interest in discussions and the respect she held among her colleagues thanks to her knowledge and previous university experience (cf. Foglarová 2005, 35).

[9]At the first meeting in 1945, the chair of the post-war *Jednota filosofická* Josef Král described the reasons for the war interruption of lectures: 'We stopped all lecturing activities. How could we look for truth about the most fundamental questions of life and existence in public lectures and discussions in a regime built upon lies and fraud, requiring absolute freedom while the most appalling regime of oppression extorted hypocrisy or at best deafening silence by terror and violence and suppressed calls for justice in prison dungeons and on the scaffolds?' (Zouhar et al. 2006, 12)

At the same time, in the protectorate, Popelová was one of the Czech thinkers who returned to the heritage of ancient philosophy. She became the chief of the *Antická knihovna* [Ancient Library] series, participating in several translations (Plutarch, Seneca). During the war period, in *Svazky úvah a studií* [Volumes of Ideas and Studies], Popelová also published her reflections on the philosophy of history, expressing historical optimism: *Dějiny a hodnoty* [*History and Values*] (1941) and *Pravda a jistoty* [*Truth and Certainties*] (1942). According to Popelová, historical experience proves that, despite all disasters, history rewards perseverance and endeavours to embark on new constructions.

Albína Dratvová left university and worked for the Ministry of Education until she retired in 1942. Dratvová suffered from the university closures, as she was in her prime at that time; she published intensively and greatly enjoyed giving lectures. In her diary, the following entry appeared on 17 October 1939: 'I am writing on a day that the Czech educated nation has received a deep wound: today, our dear Charles University was closed by the Führer's order—for three years. Thus, I gave the last lecture yesterday—and who knows when I will be able to lecture again! I loved lecturing and preparing for my lectures painstakingly, even as regards the delivery of the lecture. My work at university was an utter delight for me.' (Dratvová 2008, 321) Despite that, she kept writing during the war and published her texts in books and journals. In Václav Petr's aforementioned series *Volumes of Ideas and Studies*, she published *Smutek vzdělanců* [*Sadness of Scholars*] in 1940; later, she wrote the works *Etika tvůrčí práce* [*Ethics of Creative Work*] (1942) and *Logika a lidé* [*Logics and People*] (1944). In this period, Dratvová's influence on Czech philosophy is evident also from the number of reviews of her war publications in the aforementioned philosophical journals. Albína Dratvová's works published during the war do not reflect responses to war events or hidden traces of patriotism, which is surprising in ethically-oriented reflections. She does not pay much attention to political events of that time in her diary either. However, one of her diary entries (20 March 1939) offers a certain explanation of her 'apolitical' approach: 'A tornado struck again across our poor country, destroying it irretrievably. I tend to be silent at the worst moments, hence I did not complain on 15 March 1939 when we were occupied by German troops either.' (Dratvová 2008, 313–314).[10] Czechoslovakia was liberated in May 1945 and the situation of Czech scholars finally changed.

[10]Dratvová only rarely commented on politics and the social situation also after the Second World War and especially after the 1948 communist coup. For more, see the following subchapter describing the period between 1945–1989.

4.3 Rise of Communist Ideology

4.3.1 Historical and Social Context

After 1945, the Communist Party became stronger in Czechoslovak politics, striving to affect even the cultural and educational spheres. In the 1946 free elections, Communists won 40% of the vote and their influence became even greater. In 1947, the first woman to join the Czechoslovak government—Ludmila Jankovcová (1897–1990), a member of the Czechoslovak Social Democratic Workers' Party, which later merged with the Communist Party of Czechoslovakia—became the Minister of Industry.

Following the liberation of the Czechoslovak Republic in 1945, the universities in Prague and Brno were reopened. In 1946, Palacký University in Olomouc was founded. German universities and other German academic institutions were shuttered by the decree of the President of the Republic on 18 October 1945. Academic research was then conducted only at Czech and Slovak institutions. War refugees returned from abroad, although some of them did not stay for long: fearing further persecution after the February 1948 Communist coup, they again left Czechoslovakia.

After 1948, Communists took control of the key positions in science and research. In the humanities, Marxism became the only acceptable opinion; other approaches were not recognised for many years. The late 1940s saw the first political trials in Czechoslovakia, culminating in the purges of the 1950s. Citizens who disagreed with the Communist regime were persecuted. Some managed to emigrate; others were sentenced to long prison terms or suffered other types of repression in their daily life. In one trial, politician and representative of the pre-war women's movement Milada Horáková (1901–1950) was executed, despite the protests of prominent foreign public figures.

In the 1960s, there was a gradual liberalisation of the regime, culminating in the reviving process of the Prague Spring of 1968. In the mid-1960s, more scholars were allowed to travel abroad and be in touch with the rest of the scientific world after a long period of relative isolation. The invasion of Czechoslovakia in August 1968 put a dramatic end to this period. The Communist regime launched the process of 'normalisation', during which noncompliant experts were again excluded from academic institutions and universities. In 1977, dissidents issued the declaration of human rights known as Charter 77, which was signed by a number of figures from contemporary Czechoslovak science and culture. The signatories of Charter 77 were persecuted and some were even condemned to prison. Many scientists were not allowed to return to their scholarly activities and re-join academic institutions until the Velvet Revolution in 1989.

Women in Czech science and philosophy between 1945 and 1989 must be introduced by an explanation of the specific status of women in the socialist order, which differs greatly from the situation in Western Europe. Likewise, the emancipatory tendencies and reception of feminist thinking at that time took place in a way that can hardly be understood without sufficient explanation (cf. Nečasová 2011; Štrbáňová 2012; Havelková and Oates-Indruchová 2014; Wagnerová 2017). During the Second

World War, women in many countries obtained many previously inaccessible jobs. The physical absence of men resulted in women's participation in production and generally in the functioning of life away from the frontlines. While a majority of women in Western Europe and the United States left these jobs after the war, women in the socialist bloc countries stayed in them. Within the socialist ideology, the woman should stand by the man's side in the working process and play the same role as he does in building socialism. However, the traditional role of women in the households and in children's upbringing was preserved, even though women were expected to be assisted by a network of institutions (nurseries, kindergartens, and after-school child care). Thus, women in socialist Czechoslovakia were forced to assume a double burden: in addition to their full-time jobs, they were supposed to take care of the household and children. From research conducted within the international project Women's Memory, it is evident that Czech women were proud of coping with this burden and the demands of their position made them more self-confident (cf. International project Women's Memory 2003). Even though they were aware of the gender inequalities related to the gender pay gap and distribution of top jobs, they did not question the issue of emancipation—in accordance with the ruling ideology, they regarded themselves as already emancipated.

Pre-war women's organisations were gradually replaced by a single institution—the Czechoslovak Women's Union, which determined the ways in which women presented themselves in the socialist regime. Women who were engaged in the pre-war women's movement met various fates. Some were involved in socialist institutions, others retired from public life or were even persecuted by the regime.

The issue of gender equality and revising the traditional view was not on the agenda in dissident movements (cf. Linková and Straková 2017). The commonly cited reason was the effort to not break up the unity of opposition movements with internal disputes. The resolution of women's uneven status was postponed for later, after democracy had been restored. Current scholars point out that women's activities in the dissent are yet to be thoroughly explored (cf. Ženy v disentu 2019). In the post-November era, women were depicted more often as helpers who prepared the ground for men—the active dissidents—and their actions have not yet been systematically scrutinised.

Even though the reflection of the foreign liberal feminist movement was not supported within the socialist order, there are still more translations of feminist texts and responses to them in Czechoslovakia in the more liberal period of the 1960s. One appropriate example is the reception of the Czech translation of *The Second Sex* by Simone de Beauvoir from 1966, especially the polemic outburst that occurred in the influential magazine *Literární noviny* [*Literary News*] in the following year (cf. Musilová 2007). This discussion between Czech philosophers shows how feminist theories were approached in the Czech milieu of the 1960s. While Jan Patočka and Irena Dubská emphasised the pertinence of Simone de Beauvoir's work and a necessary revision of the traditional view of women, Ivan Sviták refuted the text and even carried out personal attacks on the author. Other than that, a more distinct reception of liberal feminist ideas was not seen in Czechoslovakia until the 1989 Velvet Revolution.

The following subchapters introduce the types of Czech women in science and philosophy between 1945 and 1989 on the basis of their ranking in academic institutions and relation to the socialist regime. This distinction proves more suitable than a thematic categorisation, as it explains many decisions made by educated women and their approach to work. In some cases, individual attitudes were changing, especially regarding the more freethinking regime in the 1960s, the reformative movement of the Prague Spring, and the subsequent period of normalisation. It must be noted that the conditions for scientific work varied locally; institutions outside Prague were usually not under such strict supervision from the regime.

In addition to universities, women could work in archives where the conditions mostly were freer. Hence, it is possible to track the activities of female scholars related to Comeniological research. Many prominent figures in the Czech sciences dealt with the study of the works of J. A. Comenius after they were forced to leave more prestigious jobs at universities.

Most women with scientific and philosophical interests fall into the 'grey zone' that included most inhabitants of then-socialist Czechoslovakia. They were keen on doing their work without disruption, and were neither open supporters nor critics of the regime. The diary of philosopher Albína Dratvová holds an impressive description of the atmosphere of the post-war grey zone. On 15 August 1961, Dratvová noted that after 1948, she was in 'philosophical quarantine (just like everybody else who was not explicitly involved in working for the Party)'. (Dratvová 2008, 449) The term 'quarantine' aptly describes the situation of intellectuals forced to work only under the supervision of Communist institutions, with no contact with foreign 'Western' science. In some cases, women in academic institutions took an active part in the construction and re-structuring of institutions along socialist lines. Examples of this include the philosopher Jiřina Popelová, who became the rector at the university in Olomouc in 1949–1953, and the Communist politician and science populariser Luisa Landová-Štychová.

In contrast to the women who pursued careers within socialist institutions, there were women who were forced to leave Czechoslovakia, who moved on the fringes of academic research, or who directly participated in the dissident movement. These women's personal attitudes and fields of intellectual interest are often closely interrelated. For example, the work of the philosopher Božena Komárková was strongly influenced by her roots in the tradition of the Czech Brethren. Two totalitarian regimes had a dramatic impact on the life of Růžena Vacková, an archaeologist and art theoretician, who was imprisoned in the Second World War and later condemned to prison again by the Communist regime. Both Komárková and Vacková strove to continue in their scholarly work even under harsh conditions. Komárková participated in home seminars and Vacková lectured for her fellow prisoners.

4.3.2 Involvement Within Academic Institutions

This section introduces women from various disciplines who managed to have a successful career in Czechoslovakia after 1945. As stated in the Introduction, the primary focus is on the field of philosophy: only the most prominent figures from other disciplines are included here, listing their main works or fields of interest and outlining their publications.[11]

Doctors

The first large group is composed of doctors—pre-war graduates from medical faculties who worked also within the period described in the previous chapters. In many cases, their lives were complicated for political and personal reasons.

The fate of Vlasta Kálalová-di Lotti was so dramatic that it even inspired the Czech writer Ilona Borská to write a popular novel about it (cf. Borská 1978). From a young age, Vlasta Kálalová-di Lotti (1896–1971) possessed a gift for languages, although she chose to study surgery as her university major. In addition to other subjects, she learned Arabic and Turkish; after she completed her studies, she departed for the Near East. From 1925, she worked in Baghdad where she built a Czechoslovak hospital. She specialised in tropical diseases and was interested also in entomology. In 1932, she returned with her family to Czechoslovakia. She tragically lost her husband and two children at the end of the war. After the war, she lectured and travelled; however, her life and work did not become popular until the Borská's novel was published in 1978.

Eliška Klimková-Deutschová (1906–1981) was deported to the Ghetto Theresienstadt (Terezín) and later to the death camp Auschwitz (Oswiecim). After the war, which she survived, she was able to work in neurology; in 1958, she successfully defended her doctoral thesis on the topic of neurasthenia and pseudo-neurasthenia. In 1961, she was appointed a chief of the neurological clinic of the Medical Faculty of Charles University in Pilsen and became an associate professor. In April 1967, she obtained a professorship. In her research, she studied the health effects on concentration camp survivors, a situation she had herself experienced.

One of the most prominent experts in post-war Czechoslovak medicine was Helena Rašková (1913–2010), born in Zurich to a family of doctors. She graduated from the Medical Faculty in Prague in 1937 and obtained a professorship in 1957. Rašková co-established the Faculty of Children's Medical Care, founded a school of pharmacology, and engaged in experimental and clinical pharmacology. She was the first director of the Institute of Pharmacology of the Academy of Sciences in Prague. In 1963, she helped organise an international pharmacological congress in Prague and co-established the International Union of Pharmacology (IUPHAR). In the period of normalisation, she had to give up her work at the faculty and at the Czechoslovak

[11] For an overview of the 20th century Czech women working in medicine, engineering, natural science, and the humanities, see *Ženy ve vědě do roku 1945* (2013), Štrbáňová (2012), and Hoffmannová (2016).

Academy of Sciences (CSAS). She engaged in veterinary medicine and did not return to pharmacology until 1989.

Jiřina Čížková-Písařovicová (1908–1994) graduated from the Faculty of Medicine in 1931 and subsequently focused on children's medicine and endocrinology. In 1955, she was appointed a professor. She retired in 1973 and left for Canada to live with her family. She is regarded as the founder of Czechoslovak paediatric endocrinology. She was a member of the Royal Society of Medicine. For several decades, she collaborated with associate professor Bohunka Blehová (1917–1996), who focused on congenital metabolic disorders and was the first to establish first newborn screening for PKU in Czechoslovakia.

In the 1960s, more women became professors of medicine: Olga Skaličková (1906–1969) was promoted to professor of psychiatry in 1965; Věra Kadlecová (1913–2002) became a professor of ophthalmology in 1966; and Dagmar Benešová (1906–1999), a specialist in children's pathological anatomy, obtained professorship in 1968.

It is clear that many prominent Czech female doctors chose their specialisation based on their personal experience. Similarly, they were more likely to focus on disciplines concentrating on paediatrics, which is a medical field that was more easily accessible for the first female graduates from medical faculties.

Disciplines in Engineering and Natural Science

Engineering was traditionally perceived as a discipline restricted to men; hence only a few women succeeded in this field in the post-war period. Adéla Kochanovská-Němejcová (1907–1985) was the first female professor of experimental physics and the founder of X-ray structure analysis in Czechoslovakia. She studied mathematics and physics at the Faculty of Science at Charles University. At that time, she was captivated by the lectures about radiology given by Václav Dolejšek (1895–1945), under whose supervision she wrote her dissertation work focused on the study of radiation. Above all, she dealt with the application of modern physical findings in solving technical questions. Her work from 1943, *Zkoušení jemné struktury materiálu röntgenovými paprsky* [*Examining the Fine Structure of Material Through X-ray*] formed the basis for applying X-ray analysis in engineering and industry. Kochanovská published a large number of articles and studies, worked in laboratories and institutes (Institute of Spectroscopy, Department of Radiology of the Institute of Physics of Škoda Works), and lectured at universities (Charles University, Czech Technical University). Thanks to her achievements, in 1968 she became one of the first three women to become a member of the CSAS. In the press, she was referred to as the 'Czech Madame Curie'.

Another female researcher in the field of physics, Ludmila Eckertová (1924–2009), studied applied physics at the Faculty of Science at Charles University. Upon completing her studies, she served out an internship in Moscow and subsequently spent many years at the Faculty of Mathematics and Physics. She conducted research on the physics of thin films, her best-known monograph even bears the title *Fyzika tenkých vrstev* [*Physics of Thin Films*] (1973). In 1963, she became an associate professor. Her professorship was suspended at the beginning of normalisation; she was promoted to professor in 1990.

Julie Hamáčková (1892–1968) was the first woman to be appointed a professor at the Czech technical universities (1954); she was also a dean of VŠCHT (University of Chemistry and Technology) in Prague from 1957 to 1959. She is considered to be the founder of Czech hydrochemistry. She studied chemistry at ČVUT (Czechoslovak Technical University) and worked as an assistant professor at university. In 1926, she was employed in a sewage disposal plant and was the first researcher in Czechoslovakia to study the chemistry of sewage water.

The younger generation of female scientists in engineering include the first specialist in IT in Czechoslovakia, Zdena Rábová (1936–2006). Zdena Rábová studied at the Faculty of Civil Engineering of VUT (Technical University) in Brno; in the 1960s, she took an interest in programming and also enrolled in the Faculty of Electrical Engineering of VUT. At the faculty, she established a laboratory with the MSP-2A computer. In the laboratory, they created the first East-European language translator of Algol 60, and the first Czech texts about the programming language Pascal. Rábová founded information studies in Czechoslovakia and she was also the founder of the Brno school of modelling. In her scientific and research work, she concentrated on the field of simulation languages and their translators.

Humanities

The field of humanities includes Eva Kamínková (1909–1982), a classical philologist who became a professor of medieval Latin studies in Prague in 1967. In her research, she focused mainly on medieval Latin literature *Husova Betlémská kázání a jejich dvě recenze* [*Hus's Bethlehem Sermons and Their Two Reviews*] (1963). She co-authored *Slovník středověké latiny* [*Dictionary of Medieval Latin*] (1977), authored and co-authored university textbooks of Latin, and compiled several anthologies of medieval Latin texts. Another prominent classical philologist and comeniologist, Julie Nováková (1909–1991), engaged in a great number of translations and publications. She worked at the Faculty of Arts in Olomouc as an associate professor of medieval history until 1961; later, she became a researcher at the J. A. Comenius Institute for Education. She took part in the publication of several volumes of Comenius's *Opera omnia* and published her studies on Comenius in the book *Čtvrt století nad Komenským* [*Quarter of a Century of Work on Comenius*] (1990).

Another acknowledged expert was Emanuela Nohejlová-Prátová (1900–1995), who earned a degree in history in Prague. From 1930, she worked in the department of numismatics in the National Museum. She and Gustav Skalský (1891–1956) are regarded as the founders of modern Czech numismatics. After the war, Nohejlová-Prátová taught at universities in Prague and Brno; she became a university professor in 1964. In her works, such as *Základy numismatiky* [*Introduction to Numismatics*] (1975) and *Das Münzwesen Albrechts von Wallenstein* (1969), she explored older Czech coining and numismatic metrology.

The literary historian and theoretician Růžena Grebeníčková (1925–1997) dealt with Czech and world literature; she also wrote philosophical works. She earned her doctorate in Prague in 1961 and worked in CSAS. She published some of her texts and translations under the names of other authors, which was a common path to publication in the socialist period, used by intellectuals expelled from university or socially excluded. From the 1980s, Růžena Grebeníčková concentrated on body and corporality in philosophy and art; to this topic, she devoted her work *Tělo a tělesnost v novověkém myšlení* [*Body and Corporality in Modern Thinking*] (1997).

One of the most prominent experts in humanities after the Second World War was the historian and Byzantologist Milada Paulová (1891–1970), whose life and works were depicted in the previous subchapter. In the post-war period, she dealt with the history of Southeast Europe and Byzantology and worked as a professor at the Faculty of Arts at Charles University until 1961. She also engaged in editing and publishing—her work for the prestigious journal *Byzantinoslavica* is of great importance. Another Byzantologist was the philologist Růžena Dostálová (1924–2014), who made a major contribution to the foundation of Modern Greek as an independent study field at universities in Prague and Brno. For many years, she worked in the Cabinet for Greek, Roman, and Latin Studies of CSAS; she was appointed a professor after 1989.

In the field of humanities are also women who worked as editors in publishing houses of scholarly works. Working there, they could partly affect the publishing plan and push through the publications of crucial and influential texts, both originals and translations. In the 1960s, such female editors included Jiřina Zumrová in the publishing house Odeon and Eva Formánková in Mladá fronta.

Philosophy

In post-war Czechoslovakia, women intending to work in the branch of philosophy had several options.[12] One option was to work in university departments (Prague, Brno, and later Olomouc). After 1945, research in the Czech lands was conducted at universities and also within the Royal Bohemian Society for Sciences (KČSN, founded in 1784) and the Czech Academy of Sciences and Arts (ČAVU, established in 1890 under the name Emperor Franz Joseph Czech Academy for Sciences, Literature and Arts). In 1923, ČAVU started to grant regular membership to women—the first woman to be selected was the writer Eliška Krásnohorská in 1924; KČSN did not admit women until the last stage of its existence when it granted special membership to women: historian Milada Paulová, literary historian Flora Kleinschnitzerová, and philosopher Albína Dratvová.[13] After the war, the specialised departments of ČAVU were transformed into scientific institutes. After 1950, a reorganisation of Czechoslovak science was prepared, reflected in the CSAS (Czechoslovak Academy of Sciences), established in 1952. It comprised all the aforementioned institutes as well as others. The first female correspondent members of CSAS were appointed in 1968: physicist Adéla Kochanovská-Němejcová, biochemist Zora Šormová, and pharmacologist Helena Rašková. Other female correspondents and scholars followed

[12]The main resource for the following subchapters is *Slovník českých filozofů* [*Dictionary of Czech Philosophers*], from which basic facts about the life and work of the individual female philosophers have been adapted (cf. Gabriel et al. 1998).

[13]Cf. the subchapter 'In the Spirit of the First Republic'.

in the 1970s and 1980s (cf. Štrbánová et al. 2004). Some researchers who were forced to leave their university departments after the 1948 coup found work in CSAS. For the field of philosophy, the Institute of Comenius Studies at CSAS turned out to be a suitable workplace for experts who had to or wanted to work under less strict supervision from the Communist Party of Czechoslovakia. Many experts worked there who had had to stop teaching at universities, including Jan Patočka. Initially, the Comenius studies group came under the CSAS Cabinet of Pedagogy; in 1957, it was renamed the J. A. Comenius Institute of Pedagogy CSAS.[14] Many female Czech philosophers also focused on Comenius studies in the 1950s and 1960s.

Jiřina Popelová and Albína Dratvová

The following subchapter focuses on two prominent figures in 20th century Czech philosophy: Albína Dratvová and Jiřina Popelová. The life and work of Albína Dratvová is covered within multiple subchapters of this text—her pre-war career has already been introduced. This subchapter concerns Dratvová's war- and post-war work. Her life can be understood as an example of an educated woman who did not get involved in institutional networks in the socialist regime and who found herself in the so-called grey zone, and later in voluntary intellectual isolation.

Albína Dratvová received a degree in philosophy as well as physics and mathematics from Charles University. In 1932, she became an associate professor after defending her paper *Problém kausality ve fyzice* and started to lecture as an adjunct associate professor at the Faculty of Science at Charles University. Albína Dratvová stands out among other female Czech philosophers in a number of activities. She was the first woman to embark on an academic career in Czech philosophy, to become an associate professor in philosophy, and also to publish a philosophical monograph in Czechoslovakia. She was an acknowledged expert in her field, yet she failed to obtain professorship from Charles University after the war and was not allowed to continue her work in philosophy after 1948. Her last book, *Hledání ztraceného kosmu* [*In Search of a Lost Cosmos*] (1948), ran up against 'communist philosophers and "philosophers" and met with harsh criticism', as she noted in her *Scientific Diary*[15] on 6 September 1948. (Dratvová 2008, 414) Dratvová considered herself to be an 'apolitical person', which was viewed as insufficient for a university career in socialist Czechoslovakia. In her diary, she expressed her search for free academic work unlimited by social factors. During the 1950s, Dratvová was able to publish at least some articles in Czech journals, but in the early 1960s, she became so frustrated by her academic isolation that she decided to put a definite end to her philosophical work and retreat from the Czech intellectual life. She died in a retirement home in 1969.

[14]In the 1800s and early 1900s, the Czech Comenius studies focused on Comenius's pedagogic works; his philosophy was not paid much attention. This changed with authors such as J. Tvrdý, V. Hoppe and T. G. Masaryk, who were among the Czech interwar philosophers who dealt with the philosophical aspects of Comenius's works.

[15]The title of the diary in English is a reference to Faraday's *Scientific Diary*.

Most of all, Dratvová was well known and acknowledged for her works from philosophy of exact sciences. From the beginning of her career, she was also interested in ethics (her dissertation was devoted to Descartes' ethics), although she did not start to publish studies on ethical topics until the late 1930s and after the Second World War. She started to write publications with an ethical focus after the closure of Czech universities in autumn 1939 and again after she had quit her work as a clerk in the Ministry of Education and Public Enlightenment in 1942. Some of the aforementioned publications were written based on Albína Dratvová's popularisation activities. In 1939–1948, Dratvová gave regular lectures in her radio show 'Půlhodina pro ženy' ('Half an Hour for Women'). In them, she contemplated the status of women, their thinking and activities, children's upbringing, and ethical and social topics. Based on these lectures, she published *Etika tvůrčí práce* (1942) and *Duše dnešní ženy* [*Soul of Today's Woman*] (1947).

In the first ethics publication, *Smutek vzdělanců* (1940), Dratvová looked for the nature and causes of the specific sadness of scholars (scientists, teachers, and—surprisingly—clerks, rather than artists). Dratvová believed that scholars tend towards sadness due to the imbalance between their rationality, sensibility, and will; through their rational abilities, scholars particularly tend towards intellectualism, over-criticism, and self-criticism. They feel instability and scepticism towards their own capacities and the capacities of humanity in general. At the same time, they have problems in the social sphere, because it is hard for them to overcome individualism and maintain social relations. What is the remedy, according to Dratvová? She advised systematic work for the society, which can be helpful to scholars as well, through the recognition of their work and its impact. Dratvová's interest in psychology is also seen in her texts devoted to the role and situation of contemporary women—in her time, the most popular was her work *Duše dnešní ženy* (1947), based on a set of lectures for radio broadcasts. Even though Dratvová stresses that it is essential to involve women in the development of society, her view of women and their character reflects many stereotypes. In her texts, Dratvová develops a typology of modern women and characterises their qualities and weaknesses. She notes, nevertheless, that there are not strict boundaries between the types and some women can show traits of different types. Dratvová's main criteria for the types of women are their specific talents and interests. She stems from the main distinction between pyknics, organisers, and creative women. The pyknics are rather passive, with a calm and kind-hearted approach to other people. They can solve concrete problems, but they cannot deal with general solutions and are not ambitious. Women-organisers are, on the other hand, educated, they work on their own, are realistic and practical, and are not interested in romantic and sentimental behaviour. They are active proponents of women's rights. Finally, creative women are individualists, equal to men in the most important positions, e.g. management of firms, legislation, and all intellectual activities. Their uniqueness can be observed even in their childhood; they are independent in their thinking and can never adjust completely to society (there are 'masculine' elements in their behaviour). Surprisingly, Dratvová's description of the situation of women shares some common points with the analysis by Simone de Beauvoir in *The Second Sex* (1949), e.g. the study of women's creative capacities and the unsentimental approach to the characterisation of old age. While de Beauvoir uses her description to criticise and undermine the traditional scheme, Dratvová only describes the situations and accepts them as natural facts. Dratvová did not share her French colleague's radical approach and drive to change the position of women in contemporary society.

However, there is a paragraph in the *Soul of Today's Woman* where Dratvová expresses her doubts about the essentialist view of women: 'Many flaws of women's thinking were corrected by the education of women and by their work in responsible positions where all imprecision shows in the outcome of the work. It seems anyway that many defects of their lifestyle are directly formed by the way of life women were bound to for centuries; they are closer to life and its ordinary practice, but they are far from the abstract and technical world, from construction and control of the most subtle and complicated machines. The adequacy of this consideration can be, however, decided only by long experience, based on precise observation and statistics that currently cannot make use of sufficient data.' (Dratvová 1947, 112) Again, this can be compared with *The Second Sex* by de Beauvoir, where the French philosopher accumulates many historical examples to support her feminist cause. Dratvová does not reflect on de Beauvoir's work and probably was not familiar with her texts, which is regrettable.

In her study of modern women and their approach to happiness, Dratvová thus maintains the traditional view of women as creatures controlled by their emotions and dreams. Although Dratvová was successful in the field of philosophy of science and mathematics, she notes in her text that women do not tend to use abstract thinking and are focused on particular aspects of problems (cf. Dratvová 1947, 13). It is as the 'creative women' type that Dratvová would situate herself. As noted, the 'creative women' are seen as exceptions in society—that corresponds well with Dratvová's self-description in her *Diary*. In the *Diary,* Dratvová notes that her own approach is exceptional—she characterises her own thinking as 'masculine'. (Dratvová 1947, 79) She expresses her desire to fulfil intellectual ambitions and projects, in spite of social and ideological limitations. Dratvová describes intellectual work as a 'feast'—a moment of clarity when the right idea or formulation is found, and she emphasises her own creativity. In her *Diary*, she sees intellectual and creative work as the aim and fulfilment of her life. Dratvová gets frustrated when she cannot pursue her academic career or is forced to waste her time on various administrative tasks. She considers herself condemned to 'perpetual work', life spent with her books and articles, a remedy for intellectual anxiety and sadness.

How can Dratvová's paradoxical approach be explained? It is clear that she accepted the traditional approach to the roles of men and women and its distinction between 'masculine' and 'feminine' characteristics. On the other hand, she considered herself to be an exception and found pride in her uniqueness. She mentions that her mind is 'masculine', refuses to have a family, and talks about her publications as her 'true children'. Her descriptions in the typology of women (related to the 'creative women') show that she was aware of the capacities of women-intellectuals and supported their decisions. But above all, Dratvová supported the independence of intellectual life and creativity. As she notes in her *Diary*, she does not care so much about everyday life, or the theories of other philosophers—she confesses 'to live her best moments not when she is listening to the thoughts of others, but when she is thinking on her own' (Dratvová 1947, 345). Albína Dratvová is an example of a successful and original woman philosopher whose career was negatively affected by external social factors, especially the Second World War and the socialist regime in Czechoslovakia after 1948.

The most important position within post-war academic institutions was achieved by Jiřina Popelová (1904–1985). Popelová studied philosophy and classical philology at the Faculty of Arts at Charles University. In the 1930s, she wrote papers on philosophical topics about staged plays, news from literature, and reflections on the arts and works of arts. In 1929, she defended her doctoral thesis entitled *Quo modo Lucretius Epicuri de natura deorum doctrinam explicaverit*. From 1932 to 1934, she went on a study stay to Italy, during which she obtained a doctorate at the Faculty of Arts of the University of Rome. After her studies, she worked as a secondary school teacher.[16] She began her university teaching career after the war: in 1946, she became an associate professor in Brno at the Faculty of Arts at Masaryk University; in 1948, she was appointed a full professor of philosophy at the Faculty of Education at Charles University. In 1948, she became the dean of Faculty of Education of the restored university in Olomouc and from 1949 to 1953, she served as its rector. From 1953, she lectured on the history of philosophy and ethics at the Faculty of Arts at Charles University until she was sent into retirement in the early 1970s. The period in which she worked as a rector in Olomouc is rather controversial; on the one hand, Popelová was seen as a representative of the repressive regime, which severely affected the functioning of the Olomouc university and forced many prominent scholars to leave. On the other hand, her contemporaries recall Popelová's personal interest in many colleagues and students whom she was willing to provide with generous help, as well as her criticism of the normalisation development in Czechoslovakia and her strident condemnation of Jan Patočka's persecution (cf. Foglarová 2005; Gabriel 1984).

Two of Jiřina Popelová's major works were *Poznání kulturní skutečnosti* [*Insights into Cultural Reality*] (1936) and *Tři studie z filozofie dějin* [*Three Studies from the Philosophy of History*] (1947). The first is conceived as an 'introduction to noetics and the methodology of cultural studies'; the latter deals with the history of philosophy of Ancient Greek and Roman thinking, historical relativism and the possibility of overcoming it, and the perception of time in history. Both of these works sparked interest in the philosophy of history in Czechoslovakia. Contrary to historical relativism (Dilthey, Troeltsch, etc.), Popelová believed that the need for a constant revision of theoretical conceptions did not have to lead to noetic scepticism. In this respect, she also referred to Marxist philosophy (dialectical materialism), which she saw already in the mid-1930s as the most powerful attempt at historical philosophy (cf. Popelová 1936, 1947). The aforementioned books were linked by works published in the edition *Svazky úvah a studií* [Volumes of Ideas and Studies]: *Dějiny a hodnoty* [*History and Values*] (1941) and *Pravda a jistoty* [*Truth and Certainties*] (1942). The war years resonate in the *Rozjímání o českých dějinách* [*Contemplation of Czech*

[16]For Popelová's activities during the Second World War see the Sect. 4.2.

History] (1948): secret night talks held by students and their history professor on the grammar school roof offer an opportunity to reflect upon the relationship of the individual and the nation, loyalty, and work for the nation.

Before the war, Popelová was committed to social democracy, but she was one of the few philosophers who paid more consistent attention to Marxist philosophy and explored it to look for problem solving stimuli. After the war, she first acknowledged Marxist philosophy (and Marxism as such) in the brochure *Socialistický světový názor* [*Socialist Worldview*] (1946). She saw its contribution particularly in dialectical materialism (overcoming agnosticism) and in its effort to perceive humanism more profoundly. In her book *K filozofické problematice Marxova Kapitálu* [*On the Philosophy of Marx's Capital*] (1954), she highlighted the philosophical and methodological impact of Marx's *Economic and Philosophic Manuscripts*. She wanted to place the ethics in Marxist philosophy where it traditionally belongs in philosophy as such. Her *Ethics* (1962), defended as a thesis to the doctorate at Lomonosov University in Moscow) concentrates on the history of ethics as well as on contemporary ethical issues. However, systematic parts of the work are affected by the Marxist rhetoric of her time.

Jiřina Popelová later dealt with Nietzsche; she concluded her work *Rozpad klasické filosofie* [*The Decline of Classical Philosophy*] (1968) by interpreting his philosophy. As a historian of philosophy, she depicted Nietzsche's life, periodised his works, and explained the contents of his individual stages. As with other authors, she looked for a system in his works, which she found in Nietzsche's concept of will to power. Popelová provided a critique of Nietzsche's philosophy. From this perspective, she also addressed literature on Nietzsche and observed the multidimensionality of the individual interpretations. In particular, she noticed the phenomenological interpretation of Nietzsche's works in Eugen Fink and Martin Heidegger on the one hand and the critique of the decline of irrationalism in Georg Lukács's *Destruction of Reason* and Herbert Marcuse's *Reason and Revolution* on the other (cf. Popelová 1968). From the mid-1950s, Popelová dealt above all with the history of Czech and European philosophy. In her book *Jana Amose Komenského cesta k všenápravě* [*J. A. Comenius's Path to Panorthosia*] (1958), she was one of the first authors in Czechoslovakia to undertake the task of conducting an integral philosophical analysis of Comenius's thoughts and proving that his philosophical and religious ideas are the basis for all his other theoretical and practical activities. Popelová also returned to this topic in other articles on Comenius, finally even in her comprehensive monograph *Filozofia J. A. Komenského* [*Philosophy of J. A. Comenius*] (1985). Through her interest and articles on Comenius's philosophy, Jiřina Popelová made a major contribution to the development of (not only) the Czech studies of Comenius.

Comenius Studies and Czech Philosophy

Another scholar studying the works of Jan Amos Comenius was Vlasta Tatjána Miškovská (1908–1980), who had a degree in philosophy, French, and Czech from the Faculty of Arts at Charles University. She completed her doctorate in 1934 based on her rigorous work *Filosofický význam druhého vydání Kantovy Kritiky čistého rozumu* [*Philosophical Importance of the Second Edition of Kant's Critique of Pure*

Reason] (publ. in 1937). From 1949 to 1953, she worked at the T. G. Masaryk Institute; in 1954, she joined the section for Comenius studies in the Cabinet of Pedagogy of the Czechoslovak Academy of Sciences (known as the J. A. Comenius Institute of Pedagogy CSAS after 1957), where she systematically dealt with Comenius's thoughts in the European context. She participated in the translation of *Newest Method of Languages* (1964) and other works that were published in Czech in the series 'Selected works of J. A. Comenius' and she also participated in the publication of the text and commentary of the collection *Opera didactica omnia* (1957). With Jaromír Červenka, she edited the text of the first edition of the collection *De rerum humanarum emendatione consultatio catholica* [*General Consultation on the Improvement of Human Affairs*] (1966). She found her research topic in the conception of language in Comenius. She analysed it above all from the philosophical perspective; she was captivated in particular by the fifth part of the *General Consultation—Panglottia.* Based on this work, she explained Comenius's attitude to national languages and a common philosophical tongue, which he introduced as part of a wider philosophical-social system. She proved that in his construction of pansophic language, Comenius combined a rationalist perspective with an empirical one. Thanks to her appraisal of the philosophy of language in Comenius's conception, she played an irreplaceable role in the Czech studies of Comenius.

Miškovská was also well acquainted with the French philosophy of that time. In the second half of the 1940s, she took a critical part in the debate on Sartre's existentialism. In the journal *Česká mysl* in her article *Kolem existencialismu* ['On Existentialism'] (1947) she first dealt with Sartre's work *Existentialism Is a Humanism.* In 1948, Miškovská published a shorter study called *Existencialismus není humanismus* [*Existentialism Is Not a Humanism*] where she argues against Sartre's conception, as is evident from her title. First, Miškovská criticises the vague notions existentialism operates with and the severe lack of methodology in this approach: 'There is nearly no more regard to methodology and hence also a prerequisite to self-discipline in interpreting somebody else's work and in looking at the cultural life at all.' (Miškovská 1948, 14) Then she concentrates on the criticism of Sartre's view of humanism. She criticises his attempts to develop an ethical conception based on the rejection of determination and in the absence of the theory of values. In her definition of humanism, Miškovská emphasises the duties one has to another person and his or her positive acceptance. She sees nothing like that in Sartre's conception, in which the notion of 'humanism' is only to 'additionally gild the ineptness to speculate, bad literary taste, decadent annoyance of a person that does not believe in anything, does not serve anything, and spreads harmful subjectivism' (Miškovská 1948, 55). Later, Miškovská predominantly dealt with the works of J. A. Comenius, editing his texts, and translating from French and Latin. In 1965, she retired and assisted her husband, philosopher Jan Blahoslav Kozák, in his work.

Jaroslava Pešková (1929–2006) studied philosophy, history, and pedagogy at the Faculty of Arts at Charles University. From 1953, she worked in the Department of Philosophy at the Faculty of Arts at Charles University where she obtained an associate professorship in the field of the history of philosophy in 1965. From the mid-1960s, her philosophical orientation was affected by Jan Patočka's seminars on phenomenology. In the 1980s, she was forced to switch to the department of pedagogy of the Faculty of Arts at Charles University. In 1991, she was appointed a professor of philosophy with a focus on the philosophy of education. During her university studies, she began the examination of what was to be her lifelong topic—the history of Czech philosophy. In her publications at home and abroad, she dealt with the influence of philosophy on the formation of modern Czech society, Josef Jungmann, Czech Herbartism, František Krejčí, František Drtina, Masaryk, Popelová and especially Comenius. Her other research interests are related to works on Comenius: philosophy of education as a natural part of philosophy and the questions of tradition. For example, she published monographs *Utopický socialismus v Čechách v 19. století* [*Utopian Socialism in Bohemia in the 19th Century*] (1965), *Problém změny v životě člověka* [*The Problem of Change in a Person's Life*] (1985), *Problém přírody a přirozenosti v 19. století* [*The Problem of Nature and Naturalness in the 19th Century*] (1986), and *Vědomí v dějinách* [*The Role of Consciousness in History*] (1998).

Dagmar Čapková (1925–2016) studied English and Czech at the Faculty of Arts at Charles University. From 1954 to 1956, she worked as a librarian of the State Pedagogical Library in Prague where she founded a department for the study of the life and works of J. A. Comenius; from 1956 to 1992, she worked as a researcher in the department of Comenius studies and the history of pedagogy at the J. A. Comenius Institute of Pedagogy CSAS in Prague. In 1992, she became an associate professor of pedagogy at the Faculty of Arts at Charles University; in 1995, she was appointed a professor at Charles University. Čapková specialised in the works of J. A. Comenius; in Comenius studies, she also dealt with the relations of philosophy and pedagogy with regard to other fields, she examined the stages of pansophy and Comenius's view of humanity. She studied the methodological importance of Comenius's pansophic metaphysics, examined the development to the *General Consultation*, and explored Comenius's roots in the Czech and European tradition and his acceptance in the Czech lands and in Slovakia. She participated in the collective editions of Comenius's works, especially the reedition of *Opera didactica omnia* (1957), the first edition of the original of the *General Consultation* (*De rerum humanarum emendatione consultatio catholica*) (1966), the first Czech translation of the *General Consultation* (1992), the eight-volume edition *Vybrané spisy J. A. Komenského* [*J. A. Comenius's Selected Works*] (1958–1978) in Czech, and *J. A. Comenii Opera omnia* [*Works of J. A. Comenius*] (from 1969). She herself published several of Comenius's unknown works from their original Latin manuscripts and unknown letters of Comenius and his contemporaries. In the history of pedagogy and culture, she paid most attention to humanism and Enlightenment; she also studied the origins of Comenius studies starting with Palacký, Masaryk's attitude towards Czech history and his philosophy of education, and philosophy of history.

Noemi Rejchrtová (1940) graduated from the Faculty of Arts at Charles University; in 1973, she earned a doctorate in philosophy and started to work in the department of church history at the Comenius Protestant Theological Faculty (ETF) in Prague. In 1984, she became an associate professor by defending a study on the Czech utraquism of the Jagiellonian period. In 1994, she was appointed a professor and the head of the department of ETF at Charles University (authorized biography in Gabriel et al. 1998, 484). Rejchrtová specialised in the Czech reformation, the topics of the late Hussite movement (utraquism), the Calvinised Unity of the Brethren, and partly also Comenius studies—*Slovem obnovená. Čtení o reformaci*, with A. Molnár and L. Rejchert [*Restored by a Word. Readings on Reformation*] (1977); *Karel Starší ze Žerotína. Z korespondence* [*Charles Senior of Zerotin. From Correspondence*] (1982); *Jan Amos Komenský: O sobě*, with A. Molnár [*Jan Amos Comenius: About oneself*] (1987).

Olga Loužilová (1929–2018) received a degree in philosophy and Russian from the Faculty of Arts at Charles University and obtained a doctorate in philosophy and psychology. From 1960 to 1984, she worked as an assistant professor in the department of philosophy at the Faculty of Arts at Charles University. She participated in the work of the team for the history of Czech philosophy of CSAS in Prague; she focused on T. G. Masaryk's philosophy and the reception of German classical philosophy in the Czech lands in the second half of the 1800s. Loužilová sought a new perspective of Masaryk's legacy as a thinker and critically examined his topics about man. In 1967, she published her monograph on the philosophy of T. G. Masaryk *Masarykova filosofie člověka* [*Masaryk's Philosophy of Man*]; in 1970, she published a study that is utterly and intentionally independent of Marxism on the philosophy of man in Masaryk's works *Masarykův problém moderního člověka* [*Masaryk's Problem of Modern Man*]. In this monograph, she also included texts from the T. G. Masaryk archives in Prague that had not been previously published. The entire edition of this monograph was destroyed by communist authorities. The normalisation period led to the suspension of her launched habilitation and a ban on her publishing and teaching; in 1984, she was forced to take an early retirement. After her rehabilitation in 1990, she briefly returned to the Faculty of Arts at Charles University as an associate professor of the history of philosophy.

Philosophy of Science and Phenomenology

A prominent figure interested in philosophy and popularisation of science in Czechoslovakia was Luisa Landová-Štychová (1885–1969), whose life and above all political activity fall within the previous subchapter of this text. Due to her post-war interest in science, she is mentioned here as well. Luisa Landová-Štychová was a representative of the feminist and anarchist movement in the first decades of the 1900s; she later had success as a promoter of communist ideology, politician (she was elected for the parliament in 1920), and populariser of astronomy. With her husband Jaroslav Štych, she co-founded the Czech Astronomical Society in 1917 and built an observatory on Petřín. Luisa Landová-Štychová perceived astronomy as the cornerstone of the scientific worldview. After the Second World War, she was active in the Czechoslovak Astronomical Society and in the magazine *Říše hvězd* [*Empire of Stars*]. Some of her publications are: *Žena v manželství. Právní postavení čsl. občanek v manželství podle dosud plat. občan. zákonů rakouských* [*Woman in Marriage. Legal Status of Czechoslovak Women in Marriage According to the Still Valid Austrian Civil Laws*] (1923), *Výchova dětí v bezvěrecké rodině* [*Children's

Upbringing in an Atheist Family] (1947), *Astronomie v boji s Vatikánem* [*Astronomy in the Fight with Vatican*] (1951).

The next two philosophers are mentioned due to their interest in the philosophy of science and phenomenology, as an example of the broader academic focus of women working in the institutions of socialist Czechoslovakia. Marie Bayerová (1922–1997) obtained a degree in philosophy and psychology from the Faculty of Arts at Charles University; in her dissertation, she dealt with *Pojetí předmětnosti u Kanta a Husserla* [*Kant's and Husserl's Approach to Objectivity*] (1949). For a long time, she worked at the Institute of Philosophy CSAS, in the publishing department, where she devised the publication plans of the Philosophical Library for the publishing houses of CSAS and the Svoboda publishing house. She retired from the Institute of Philosophy CSAS in 1978. Beginning with her university studies, she dealt predominantly with phenomenological philosophy, especially Husserl's phenomenological method; she also translated Husserl into Czech (*Cartesian Meditations*, 1968). Another area of her interest was Bolzano's philosophy, perceived in the context of the 'genealogy' of the phenomenological method and research of German and Austrian philosophical thoughts in the Czech lands in the 1800s and 1900s. Moreover, Bayerová contributed to the study of the history of 20th century Czech philosophy (e.g. the reception and position of phenomenological philosophy in the Czech philosophical life). The mutual interrelation of these fields of interest in her work is evident especially in the monograph *Bernard Bolzano. Evropský rozměr jeho filosofického myšlení* [*Bernard Bolzano. The European Aspect of his Philosophical Thinking.*] (1994).

Jiřina Stachová (1934) completed her studies of philosophy and history at the Faculty of Arts at Charles University in 1957 and that year she became an assistant in the department of Marxism-Leninism at the Faculty of Civil and Electronic Engineering (ČVUT), where she worked until 1970 when she was hired as a researcher at the Institute of Philosophy CSAS. In 1968, she defended her dissertation *Úloha modelu ve vědeckém poznání* [*The Role of Model in Scientific Knowledge*]. While she worked at ČVUT, she was interested in the questions of natural sciences and their modern trends and discoveries. Her interest in more general questions of cybernetics led her to the topic of her dissertation work. In 1972, she co-founded the Permanent Seminar on Philosophical and Methodological Issues in Natural and Technical Sciences. In 1984, she launched a series of interdisciplinary seminars, which were immensely popular among the employees of natural-scientific and humanities institutes of CSAS and universities. For example, the first seminar was devoted to the philosophical problems of Prigogine's thermodynamics. After her 1991 retirement, she dealt with metaphor and interdisciplinary seminars (authorized biography in Gabriel et al. 1998, 534–535).

4.3.3 Emigrants and Dissidents

After the Second World War and the 1948 coup, many educated women left Czechoslovakia. The first generation of emigres included Eva Jelínková (Eve Anne

Elisabeth Reymond) (1923–1986). Eva Jelínková had a degree in ancient history from the Faculty of Arts at Charles University in Prague; from the very beginning of her studies, she was interested in Egyptology. In 1946, she emigrated to England where she started with her studies at Oxford. In her academic texts, she focused on Demotic and publishing papyrus. Another Czech emigre, writer and literary theorist Milada Součková (1899–1983), settled in the United States after the Second World War and lectured there, particularly on Czech literature (cf. Knopp 1996).

A new generation of female emigres left Czechoslovakia after 1968 and in the 1970s (cf. Štrbáňová and Kostlán 2011). In these cases, one can trace the development of their thinking and careers, which often initially developed within socialist institutions and resulted in their disagreement with the regime.

Irena Dubská (1924–2010) graduated from the Faculty of Arts at Charles University with a degree in sociology, psychology, and philosophy. From 1954 to 1974, she was a researcher at the Institute of Philosophy CSAS; in 1969 she became an associate professor of sociology; in 1963 and 1964, she was on a study visit to the Massachusetts Institute of Technology in the United States (she described the stay in her book *Americký rok* [*American Year*] (1966). In the 1960s, she became the editor-in-chief of the *Sociologický časopis* [*Czech Sociological Review*] and participated in the renewal of Czechoslovak sociology. From 1977, she lived in Austria with her husband Zdeněk Mlynář (1930–1997), a politician, lawyer, and political theorist. She focused on the topics of social and political philosophy and sociology. In her analysis of Auguste Comte's philosophy and sociology, (*A. Comte a vytvoření sociologie. K otázce povahy a zdrojů Comtova učení* [*A. Comte and the Creation of Sociology. On the Question of Nature and Sources of Comte's Teachings*], 1960), she proved that Comte's conception means a departure from the Enlightenment tradition of the 1700s as well as from the tradition of classical philosophical thinking. She wrote many newspaper articles and texts in which she popularised humanities (e.g. in *Literární noviny, Světová literatura, Vesmír, Dějiny a současnost* etc.).

Jindřiška Tichá (1937) graduated from the Faculty of Arts at Charles University with a degree in philosophy (with a specialisation in logic). From 1961 to 1968, she taught in the department of philosophy of the Faculty of Science at Charles University. Since November 1969, she has lived outside of the country: after a year in England she left for New Zealand. From 1971, she worked in the department of philosophy and department of political sciences of the University of Otago in New Zealand (authorized biography in Gabriel et al. 1998, 581). She specialised especially in political philosophy, but she has also published works from the fields of logic, epistemology, moral philosophy, and philosophy of literature. Under the name Jindra T., she wrote fiction with philosophical features, e.g. the novels *Už se neshledáme v tomto životě* [*We Will Never Meet Again in This Life*] (1993) and *Smrt a odpuštění* [*Death and Forgiveness*] (1997).

From 1962 to 1967, Gerlinda Šmausová (1940) studied sociology, culture, and adult education at the Faculty of Public Awareness and Journalism at Charles University. After August 1968, she emigrated to West Germany, where she was employed as an assistant in the department of sociology of the Faculty of Arts at the Saarland University in Saarbrücken. In 1972, she became a researcher at the Institute for Legal and Social Philosophy at the Faculty of Law and Economy at that university. In 1997, she became an associate professor of general sociology and in 2001 a professor of sociology at Masaryk University in Brno (authorized biography in Gabriel et al. 1998, 569). Gerlinda Šmausová has published a number of works in the fields of critical criminology, sociology, and feminist and gender theory, e.g. *Der Kulturarbeiter der ČSSR. Versuch einer Berufsrollenanalyse* (1974). Within her research activities, she has participated in a number of international congresses and conferences devoted to social and legal philosophy.

Female philosophers-dissidents in Czechoslovakia found themselves in a difficult situation, as they condemned themselves to academic isolation due to their attitude to the socialist regime. They often had to give up their jobs in academic institutions; in some cases, they were in danger of being arrested (cf. Linková and Straková 2017).

Irena Šnebergová (Michňáková) (1930) graduated with a degree in philosophy from the Faculty of Arts at Charles University in 1954. After defending her dissertation in the field of the history of Czech philosophical thinking in the 1800s, she worked in the Institute of Philosophy CSAS as a researcher in the department of the history of philosophy. In the 1970s, she was dismissed from the Institute of Philosophy for political reasons and subsequently employed in the administration of the Thomayer hospital in Prague where she worked for nearly 15 years. The period after her departure from the Institute of Philosophy in the 1970s is characterised by her intensive encounters with Jan Patočka (seminar on Heidegger, copies of his works, etc.) and with the circle around Karel Kosík. In 1990, she was rehabilitated and re-employed in the Institute of Philosophy CSAS (authorized biography in Gabriel et al. 1998, 569–570).

Eva Formánková (1931) studied Czech and philosophy at the Faculty of Arts at Charles University. During her studies, she started to work in the Mladá fronta publishing house as an editor. In 1966, Formánková spent several months in the United States, where she purchased a considerable volume of literature that was not available in Czechoslovakia at that time. She contributed to the publishing of texts from non-Marxist philosophy. The new series *Váhy* offered translations of authors that had been nearly unknown to the broader cultural community for twenty years: Søren Kierkegaard, Martin Buber, José Ortega y Gasset, Alfred North Whitehead; she also prepared works of other authors. However, the series was published in Mladá fronta only until 1970. After the series was banned, Formánková worked on the editorial board of translated fiction until 1974 when she was asked to quit by the new management. She stayed in personal touch with Mladá fronta until February 1977, when she signed Charter 77. She returned to Mladá fronta in November 1989 and launched new activity with the edition of Patočka's work *Tři studie o Masarykovi* [*Three Studies on Masaryk*] (authorized biography in Gabriel et al. 1998, 136–137). In the *Váhy* series, Formánková also wrote epilogues—e.g. to Kierkegaard and to Gabriel Marcel.

Peluška Bendlová (1931) studied philosophy and history at the Faculty of Arts at Charles University. In 1965, she defended there her dissertation entitled *Kapitoly z dějin katolické filozofie* [*Chapters from the History of Catholic Philosophy*]. Her main research interest was focused on neo-Thomism, Teilhardism, and Christian personalism. From 1966 to 1967, she studied as a research fellow of the French government at École Pratique des Hautes Études in Paris. With the arrival of normalisation, she was fired from CSAS. There was a ban imposed on her publication on Thomistic metaphysics and natural theology called *Bůh a svět* [*God and World*]. From 1970 to 1990, she had no permanent job; sometimes she interpreted from Polish and occasionally also from French. She returned to the Institute of Philosophy in 1990, with a focus on Christian and Jewish philosophy of dialogue (Gabriel Marcel, Martin Buber, Emmanuel Lévinas, etc.) and on modern French thinking. Bendlová is also a translator of philosophical and sociological literature from French and Polish (authorized biography in Gabriel et al. 1998, 32–33).

In the 1920s, Božena Komárková (1903–1997) studied philosophy, history, and geography at the Faculty of Arts at Masaryk University. Her teachers were direct followers of T. G. Masaryk: philosopher Josef Tvrdý, historian Julius Glücklich, and sociologist and philosopher Inocenc Arnošt Bláha. She was also influenced by the local association of the YMCA, a Christian student movement, and the leading figures connected to it. Komárková first taught at primary schools and later at grammar schools in Ivančice and Ostrava. At the very beginning of the Nazi occupation, she engaged in the resistance organisation Defence of the Nation. She was arrested in January 1940 and after a two-year detention in Wroclaw, she was sentenced to twelve years in prison 'for plotting treason'. After the liberation, she taught history and philosophy at a classical grammar school in Brno and started to write the dissertation *Obec Platónova a Augustinova* [*The City of Plato and Augustine*], which she submitted to the Faculty of Arts at Masaryk University and received the degree of a Doctor of Philosophy in March 1948. She continued with a habilitation work called *Lidská práva ve filosofii XIX. století* [*Human Rights in 19th Century Philosophy*]; however, she was no longer allowed to submit it. She had to wait until June 1992 to be able to defend it and obtain an associate professorship. For ideological reasons, she was not allowed to work in education after 1950. She had several jobs and retired in 1951. Nevertheless, she continued to teach and educate others: after the closure of the YMCA in the 1950s, she lectured in the Evangelical Church of the Czech Brethren and in her flat. She entered the dissent movement as a signatory of Charter 77; she was interrogated and her flat was searched by the police in 1977 and 1988. She later contributed to samizdat series and anthologies; a number of her studies were published abroad. In them, she promoted Christian humanism, interpreted the development of modern society in a Christian way, and emphasised human rights. These works include *Sekularizovaný svět a evangelium* [*Secularised World and Gospels*] (Zurich 1981; 1992), *Původ a význam lidských práv* [*Origin and Importance of Human Rights*] (1986; the 1990 extended edition also includes her doctoral dissertation), *Božena Komárková a její hosté* [*Božena Komárková and Her Guests*] (1980; 1991). Her personal courage was awarded; in 1991, President Václav Havel bestowed upon her the Order of T. G. Masaryk for her lifelong struggle for democracy.

In her texts, Božena Komárková addressed a fundamental question: Are human rights only a transient product of one epoch or are they the permanent property of our civilisation for which we are responsible? Her answer reflects her belief that human rights are a fundamental precondition of life in modern society. Her philosophical reflection starts with exploring the roots and the foundation of the idea of human rights, which is found in reformation thinking, especially in Calvin, and in the natural law theories of Enlightenment thinkers. Komárková asserts that there are two traditions in modern-day humanism. One is the tradition of the Anglo-Saxon thinking of responsibility and freedom of human character (John Milton, John Locke, John Stuart Mill), which was included in the Declaration of Human Rights in the United States (1776). The other tradition is continental thinking (Rousseau, Hegel, Comte, Marx, and others), which makes collectivity absolute (general will, state). That results in tragic consequences and totalitarian regimes. From the continental tradition, Komárková defends only Kant and Masaryk (cf. Komárková 1990).

Among female scholars-dissidents and emigrants, one outstanding story is the tragic fate of the classical archaeologist and arts theoretician Růžena Vacková, who was one of the first pre-war associate professors and was destined for an excellent academic career (cf. Vacková 1999). Růžena Vacková (1901–1982) was born in Velké Meziříčí; she completed her secondary school studies in Vyškov and Brno. She studied at the Faculty of Arts in Prague and became a student of art history professor Vojtěch Birnbaum, a representative of the Viennese school. Vacková obtained associate professorship in 1930 by defending a work on Roman historical reliefs. In 1932, she published a work called *Lanna a antika* [*Lanna and Antiquity*], in which she studied terracotta from the collection of the founder of the Museum of Decorative Arts, Vojtěch Lanna. In addition to works from classical archaeology, she studied aesthetics and wrote studies about the theatre (the collection of texts *Výtvarný projev v dramatickém umění* [*Artistic Expression in Performing Arts*] came out in 1948). During the war, she participated in the resistance and was imprisoned in Prague. In 1947, she was appointed a professor of classical archaeology at Charles University; however, her academic career was drastically suspended soon after that. After the war, Vacková held meetings for young Catholics, and after 1948 she openly stood up for expelled students. She had to leave university in 1950; she was condemned to 22 years in prison for her 'anti-state' activity in 1952 and released in 1967. In prison, she lectured for her fellow-prisoners; improvised lectures from the notes of one of her students were published in 1999 as *Vězeňské přednášky* [*Lectures from Prison*]. After her release from prison, Vacková organised home seminars and was one of the first female signatories of Charter 77. When she was set free, she commenced her key work devoted to the style of art, in which she developed the ideas of the Viennese school and which she was preparing for publication with her friend Josef Zvěřina. However, it was not possible to publish the book before 1989 (cf. Vacková 1993).

References

Baborová MZ (1902) Úvod ke studiu o tukovém tělese členovců: (de corpore adiposo Arthropodum). Zdeňka Baborová, Praha

Bahenská M, Musilová D, Heczková L (2011) Iluze spásy: české feministické myšlení 19. a 20. století. Veduta, České Budějovice

Bartlová M, Pachmanová M (eds) (2008) Artemis a dr. Faust. Ženy v českých a slovenských dějinách umění. Academia, Praha

Bečvář J et al (1998) Dějiny Univerzity Karlovy 4, 1918–1990. Karolinum, Praha

Borská I (1978) Doktorka z domu trubačů. Mladá fronta, Praha

Brádlerová D (2018) Czechoslovak National Research Council, 1924–1953. In: Míšková A, Franc M, Kostlán A (eds) Bohemia docta: the historical roots of science and scholarship in the Czech lands. Academia, Praha, pp 365–393

Čáda F (1911) Účast žen ve filosofii antické. Listy filologické/Folia philologica 38:1–10, 81–93, 178–201, 321–328

Dratvová A (1921) O stavech podvědomých a Freudově psychoanalyse. B. Kočí, Praha

Dratvová A (1924) Positivismus ve fysice. Jednota matematiků a fysiků, Praha

Dratvová A (1925) O tajích podvědomí. Jan Košatka, Praha

Dratvová A (1931) Problém kausality ve fysice. Česká akademie věd a umění, Praha

Dratvová A (1934) Heuristické předpoklady fysikálního badání. Přírodovědecká fakulta, Praha

Dratvová A (1939) Filosofie a přírodovědecké poznání. Česká grafická unie, Praha

Dratvová A (1940) Smutek vzdělanců. Václav Petr, Praha

Dratvová A (1947) Duše dnešní ženy. Kompas, Praha

Dratvová A (2008) In: Čápová KA, Heczková L, Leštinová Z (eds) Deník 1921–1961: scientific diary. Academia, Praha

Fabiánová M (1898) O objevu Zeemanově. Časopis pro pěstování matematiky a fysiky 27:199–203

Foglarová E (ed) (2005) Filosofka Jiřina Popelová. Univerzita Karlova, Praha

Gabriel J (ed) (1984) Jiřina Popelová, filozofka a učitelka filozofie. UJEP, Brno

Gabriel J, Hroch J, Nový L (eds) (1994) Czech philosophy in the XXth century. Council for research in values and philosophy, Washington

Gabriel J, Nový L, Zouhar J (1995) Česká filozofie ve 20. století. I., Směry, osobnosti, problémy. Vydavatelství Masarykovy univerzity, Brno

Gabriel J et al (eds) (1998) Slovník českých filozofů. Masarykova univerzita, Brno

Gabriel J, Pavlincová H, Zouhar J (eds) (2001) Česká filosofie mezi dvěma světovými válkami: náčrt. Masarykova univerzita, Brno

Harna J (2009) First Czechoslovak Republic (1918–1938). In: Pánek J, Tůma O et al (eds) A history of the Czech lands. Karolinum, Prague, pp 393–432

Havelková H, Oates-Indruchová L (eds) (2014) The politics of gender culture under state socialism: an expropriated voice. Routledge, London

Havelková M (2012) Ženy v české vědě za druhé světové války. Dissertation thesis. Západočeská univerzita v Plzni, Plzeň

Havlíková L (2009) První žena na Universitě Karlově (Paměti M. Paulové). Slovanské historické studie 34:127–165

Havlůjová H (2005) Okouzlení Egyptem: Ludmila Matiegková (1889–1960). Set Out, Praha

Hoffmannová J (2016) Prvenství žen. Ženy iniciativní, vzdělané a tvořivé. Ústav T. G. Masaryka, Praha

Honzík JM (2010) Ing. Zdena Rábová, CSc. (1936–2006). Zkus IT. http://www.zkusit.cz/zeny-it/historie-zdena-rabova.php. Accessed 14 Jan 2019

Janda J (1914) Velký ilustrovaný přírodopis všech tří říší. Díl 2. Ústřední spolek jednot učitelských, Praha

Jemelka P (2016) Reflexe environmentální problematiky v dějinách české a slovenské filosofie. Filosofia, Praha

Jůnová-Macková A (2014) Orientální ústav a československé pronikání do Orientu 1928–1938. Stipendia hospodářského odboru. Časopis Národního muzea 183:19–34
Knopp F (1996) Česká literatura v exilu 1948–1989. Makropulos, Praha
Komárková B (1990) Původ a význam lidských práv. Státní pedagogické nakladatelství, Praha
Kostlán A, Janko J, Niklíček L (2018) The organizational structure of science, 1882–1985. In: Míšková A, Franc M, Kostlán A (eds) Bohemia docta: the historical roots of science and scholarship in the Czech lands. Academia, Praha, pp 269–308
Křemenová A (2005) Anna Pammrová – životopis. Sursum, Tišnov
Lenderová M et al (eds) (2009) Žena v českých zemích od středověku do 20. století. Nakladatelství Lidové noviny, Praha
Linková M, Straková N (eds) (2017) Bytová revolta: Jak ženy dělaly disent. Academia, Praha
Martínek J (2008) Geografové v českých zemích 1800–1945: (biografický slovník). Historický ústav, Praha
Martínek J (2010) Radost z poznání nemusí vést k uznání. Julie Moschelesová. In: Vošahlíková P et al (eds) Cesty k samostatnosti: portréty žen v éře modernizace. Historický ústav, Praha, pp 176–189
Matiegková L (1916) Názory starých Egypťanů o duši. Věstník České akademie císaře Františka Josefa pro vědy, slovesnost a umění 25:47–71, 124–155
Míšková A (2002) Německá (Karlova) univerzita od Mnichova k 9. květnu 1945. Karolinum, Praha
Míšková A, Franc M, Kostlán A (eds) (2018) Bohemia docta: the historical roots of science and scholarship in the Czech lands. Academia, Praha
Miškovská VT (1947) Kolem existencialismu. Česká mysl 40:167–174
Miškovská VT (1948) Existencialismus není humanismus. Kostnická jednota, Praha
Moudrá P (1920) O potřebě reformy výchovy. Rolnická tiskárna, Brno
Moudrá P (1922) Poslání ženy ve světle theosofie. Sfinx, Praha
Moudrá P (1925) Můj odkaz světu. Karel Vačlena, Mladá Boleslav
Moudrá P (1926) Obrození duší. Karel Vačlena, Mladá Boleslav
Musilová D (2007) Na okraj jedné návštěvy: Simone de Beauvoir v Československu. Univerzita Hradec Králové, Hradec Králové
Nečasová D (2011) Buduj vlast – posílíš mír!: ženské hnutí v českých zemích 1945–1955. Matice moravská, Brno
Nikšová G (1971) Nedovolené prerušenie tehotenstva v československom trestnom práve. Slovenská akadémia vied, Bratislava
Pammrová A (1917) Alfa: embryonální pokus o řešení ženské otázky. Höschl, Klatovy
Pammrová A (1925) Cestou k zářnému cíli. Höschl, Klatovy
Pammrová A (1936) Zápisky nečitelné. Za lepším životem, Brno
Pánek J (1974) Pavla Moudrá. Poznámky k životu, působení a pozůstalosti české spisovatelky a bojovnice za mír. Středočeský sborník historický 9:215–248
Pešek J, Ledvinka V (eds) (1996) Žena v dějinách Prahy. Scriptorium, Praha
Plamínková F (ed) (1930) Masaryk a ženy: sborník k 80. narozeninám prvního presidenta republiky československé T.G. Masaryka. Díl 1. Ženská národní rada, Praha
Popelová J (1936) Poznání kulturní skutečnosti. Kruh přátel filosofie Vladimíra Hoppe, Praha
Popelová J (1947) Tři studie z filozofie dějin. Dělnické nakladatelství, Praha
Popelová J (1968) Rozpad klasické filosofie. Svoboda, Praha
Rádl E (1932) Proti tak zvané sociální indikaci: námitky proti návrhu nového zákona o umělém potratu, které autor přednesl na sjezdu Akademické YMKY v Novém Městě nad Metují dne 11. července 1932. YMCA, Praha
Rádl E (1933) O ženském hnutí. Čin, Praha
Rothkirchen L (2006) The Jews of Bohemia and Moravia: facing the Holocaust. University of Nebraska Press, Lincoln
Štrbáňová S, Stamhuis IH, Mojsejová K (eds) (2004) Women scholars and institutions, proceedings of the international conference, Prague, 8–11 June 2003. Výzkumné centrum pro dějiny vědy, Praha

Štrbáňová S, Kostlán A (eds) (2011) Sto českých vědců v exilu. Academia, Praha
Štrbáňová S (2012) Ženy ve vědě v letech 1840–1989. http://www.veda.cz/article.do?articleId=22769. Accessed 14 Jan 2019
Těšínská E, Schwippel J (2018) Masaryk Academy of Labour (MAL). In: Míšková A, Franc M, Kostlán A (eds) Bohemia docta: the historical roots of science and scholarship in the Czech lands. Academia, Praha, pp 309–364
Uhrová E (2012) Anna Honzáková a jiné dámy. E. Uhrová, Czechia
Vacková R (1993) Věda o slohu. Aula, Praha
Vacková R (1999) Vězeňské přednášky. Archiv UK v Praze, Praha
Vohlídalová M (2018) Vědkyně v době pětiletek a v době soutěže: ženské vědecké dráhy před rokem 1989 a po něm. Sociologický časopis/Czech Sociol Rev 54:3–34. https://doi.org/10.13060/00380288.2018.54.1.394
Wagnerová A (2017) Žena za socialismu. Československo 1945–1974 a reflexe vývoje před rokem 1989 a po něm. Slon, Praha
Zouhar J (1999) Studie k dějinám českého myšlení 20. století. Masarykova univerzita, Brno
Zouhar J (2008) Dějiny českého filozofického myšlení do roku 1968: stručný přehled. Academicus, Brno
Zouhar J (2009) Česká filozofie v šedesátých letech: poznámky k tématu. Academicus, Brno
Zouhar J, Pavlincová H, Gabriel J (eds) (2003) Spory v české filosofii mezi dvěma světovými válkami: výbor textů. Masarykova univerzita, Brno
Zouhar J, Pavlincová H, Gabriel J (eds) (2006) Filosofie za protektorátu: vybrané texty z českých filosofických časopisů. Masarykova univerzita, Brno
Zouhar J, Pavlincová H, Gabriel J (2007) Česká filosofie v letech protektorátu: poznámky k tématu. Masarykova univerzita, Brno
Zouhar J, Pavlincová H, Gabriel J (2013) Česká filosofie v letech 1945–1948. Academicus, Brno
The Centre for Gender and Science (2019) http://genderaveda.cz/en/gender-and-science/. Accessed 14 Jan 2019
International project Women's Memory (2003) Gender studies. http://www.womensmemory.net. Accessed 14 Jan 2019
Ženy v disentu (2019) Sociologický ústav AV ČR. http://zenyvdisentu.soc.cas.cz/literatura. Accessed 14 Jan 2019
Životopisy vědců. Techmania Science Center (2019) https://edu.techmania.cz/cs/encyklopedie/zivotopisy-vedcu. Accessed 14 Jan 2019
Ženy ve vědě do roku 1945 (2013) 3.0 Česko. http://albina.ff.cuni.cz/index.php/Hlavn%C3%AD_strana. Accessed 14 Jan 2019

Chapter 5
Conclusion

The volume presents a brief chronological overview of significant Czech women philosophers and scientists, focusing mainly on the activities of women in the 1900s. However, the first chapter was devoted to an earlier period (*Before the 1820s: Setting the Stage for Women's Intellectual Progress*) and traced the intellectual interest of women in various fields, starting with learned women in convents and religious movements to the women working on philosophical translations, and finally to the authors of manuscripts, mostly on pedagogical and ethical themes.

The main chapter (*1820s–1918: An Expansion of Women's Intellectual Pursuits*) studied the theme of the book in relation to the Czech National Revival movement, focusing on the increasing interest in Czech language and culture. The role of the National Revival process explained various choices by learned Czech women and their reactions to foreign influences and motivations. At the end of the chapter, the situation of the first Czech women at universities was described.

The final chapter (*Czech Women in Philosophy and Science since 1918*) covered the period between the two wars, witnessing the arrival of women in academic institutions, though their ascent was often complicated and long. Women also worked on academic themes out of the institutional frame, in contact with scientific institutions or as solitary thinkers. A short subchapter was devoted to the situation of Czech women scholars during the Nazi occupation and their reactions to the war atrocities. The last part of the volume described the situation after the communist coup d´état in 1948, when the position of women in academic institutions depended largely on their relation to the socialist regime. After 1948, Czech women could obtain important positions in various academic fields unless they expressed a negative attitude to the regime. Some women rejected the regime and were forced to work in the dissident movement, or they decided to leave the country to be able to work in freedom. After the Velvet Revolution in 1989, the women who had opposed the socialist regime were finally able to return to academic institutions.

Z. Jastrzembská et al., *Czech Women Philosophers and Scientists*,
SpringerBriefs in History of Science and Technology,
https://doi.org/10.1007/978-3-030-61516-1_5

The overview *Czech Women Philosophers and Scientists* offers many paths that can be explored and developed in comparison with the situation of learned women in the broader, international context. As we showed in the book, there are many sources that merit study and detailed commentary.

Historical Timeline

Historical events in the Czech lands and in the world	Women in the Czech lands
400s–500s: arrival of the Slavs in what is now the Czech Republic	Legend of Libuše
623–624: foundation of Samo's Empire, union of West Slavic tribes	
830–836: foundation of Great Moravian Empire	
863: arrival of missionaries Cyril (Constantine) and Methodius in Great Moravia	
late 800s: beginnings of the Czech state: Přemyslid duke Bořivoj converted to Christianity	
mid 900s: foundation of first convent in Bohemia (at St George's Church in Prague Castle); foundation of Diocese of Prague (973); foundation of Břevnov Monastery near Prague (993)	Mlada, founder of St George's Convent (930/935–994)
1085: Vratislaus II crowned first Czech king	
1212: Golden Bull of Sicily (heredity of royal title)	Agnes (1211–1282), the abbess of the Poor Clares Monastery 'Na Františku' in Prague Kunigunde (1265–1321), abbess of St George's Convent
1278: Battle on the Marchfeld	
1310: accession of the Luxembourg dynasty to the Czech throne	
1346–1378: reign of Charles IV	
1348: establishment of the university in Prague	
1415: execution of Jan Hus	

(continued)

Z. Jastrzembská et al., *Czech Women Philosophers and Scientists*,
SpringerBriefs in History of Science and Technology,
https://doi.org/10.1007/978-3-030-61516-1

(continued)

Historical events in the Czech lands and in the world	Women in the Czech lands
1419–1434: Hussite wars	
1434: Battle of Lipany	
1457: Unity of the Brethren founded	
1526: Habsburg dynasty on the Czech throne (Ferdinand I), until 1918	Catherina Elisabeth von Kamenek Alena Meziříčská of Lomnice (died in 1585) founds a Latin Lutheran academy (grammar school) on her estate in Velké Meziříčí in 1576 Anna Adam of Aventino (1550–1605) women from the Unity of the Brethren women missionaries Polyxena of Lobkowitz (1566–1642)
1609: Rudolf's Letter of Majesty	Elisabeth Jane Weston (1581–1612) Sophie Brahe (1559–1643), in Prague Rebecca (Rivka) bat Meir Tiktiner (died in Prague around 1550): ethical and medical treatise *Meneket Rivkah* [*Rebecca's Nursemaid*] published in Prague in 1609 Chava (Eva) Bacharach (c. 1580–1651)
1620: Battle of White Mountain, defeat of Bohemian estates	
1618–1648: Thirty Years' War	Marie Ernestine von Eggenberg, née von Schwarzenberg (1646–1719)
1774: compulsory school attendance (six years) introduced by Maria Theresa of Austria	Sporck sisters: Maria Eleonora (1687–1717)and Anna Katharina (1689–1754)
1781: Patent of Toleration issued (equality of non-Catholic religions) and abolished serfdom in the Czech lands by Joseph II	
late 1700s: beginnings of the Czech National Revival	
1805: Battle of Austerlitz	Alexandra Shuvalova (1775–1847) Karolina Ferdinandi (1777–1844)
1815: Napoleon's defeat at Waterloo; Holy Alliance; Congress of Vienna 1848: wave of revolutions in a number of European countries	1813: foundation of the first women's club in the Czech lands—Unity of Ladies and Girls to Promote Women's Education 1825: Josephine Kablick (1787–1863) becomes a member of the Institute of Plant Exchange early 1840s: group of women around Bohuslava Rajská (1817–1852) work on the *Encyclopaedia for Ladies* 1861: first publication of a fiction and fashion magazine for women called *Lada*

(continued)

(continued)

Historical events in the Czech lands and in the world	Women in the Czech lands
1866: Austro-Prussian War; Austria defeated 1867: Austro-Hungarian Compromise, Austria-Hungary established	1865: foundation of the American Ladies' Club (by Vojta Náprstek on 15 Jan 1865) as a free association and educational institution for women
	1865: foundation of the first Czech technical school for women
	1869: 1 Jan 1869 first kindergarten in Prague foundation of Physical Education Association of Prague Ladies and Girls
	1870: Imperial Royal Institute for the Education of Women Teachers (Paedagogium) foundation of Vesna, association for the education of women, in Brno
1882: division of the university in Prague into German and Czech universities	1871: foundation of Czech Women's Manufacturing Society in Prague, initiators: Karolina Světlá, Eliška Krásnohorská
	1890: Minerva, the first Czech grammar school for girls, founded in Prague
	1896: permission to recognise doctoral degrees in medicine of women who earned them abroad
	1897: admission of women as regular students of the faculties of arts of Austrian universities, from 1900 also at faculties of medicine 15–17 May 1897: first congress of Czech women in Prague
	1901: first graduations from the Faculty of Arts of the Czech university in Prague (Marie Zdeňka Baborová in zoology, Marie Fabiánová in mathematics)
	1902: first woman graduates from the Faculty of Medicine of the Czech university in Prague—Anna Honzáková
	1903: doctoral graduations of women in the fields of history, classical philology, and botany
	1905: first female motorist with a driving licence—Olga Procházková
	1907: first secondary school for girls in Moravia (Valašské Meziříčí and Brno)
	1910: first Czech professional female journalist Olga Fastrová becomes an editor of *National Policy*, the most widely read newspaper at that time

(continued)

(continued)

Historical events in the Czech lands and in the world	Women in the Czech lands
	1911: first Czech female aviator—Božena Laglerová; Czech Technical University in Prague admits women in the fields of architecture, chemistry, agriculture, insurance, and economy
1914: start of the First World War 1917: revolution in Russia 1918: defeat of Germany and Austria-Hungary and foundation of the Czechoslovak Republic; T. G. Masaryk becomes the first president	1912: in the by-elections to the Czech Diet, the first woman elected on 13 June 1912 in the Mladá Boleslav–Nymburk district—Božena Viková-Kunětická (1862–1934); however, the election was not recognised and confirmed by Franz, the Prince of Thun, the Governor of Bohemia. The election was not recognised until 1918, when she became a member of the Revolutionary National Assembly
	1918: Eight women in the new Czechoslovak parliament; first regular female students at the Faculty of Law of Charles University
	1920: Constitution of CSR includes gender equality and universal suffrage 18 April 1920: elections for the National Assembly—103 women out of 300 MPs 25 April 1920: Senate elections—three women out of 150 senators
	1921: first Czech female engineers graduate
	1922: women admitted for study at the Academy of Fine Arts
	1925: first woman to habilitate in CSR—Milada Paulová, field: general history of Southern Europe and the Balkans
	1926: first woman to habilitate at the Faculty of Medicine of the German University in Prague—Hedwig Langecker
1929: world economic crisis	1928: first woman to habilitate in legal theory—Jarmila Veselá
1933: Adolf Hitler appointed the Chancellor of Germany	1932: habilitation (associate professorship) of Albína Dratvová, first woman at the Faculty of Science of Charles University in the field of philosophy of natural sciences
	1934: first female university professors in CSR at the University of Bratislava and German University in Prague

(continued)

(continued)

Historical events in the Czech lands and in the world	Women in the Czech lands
September 1938: Munich Agreement March 1939: occupation of Czech lands by the German army: establishment of the Protectorate of Bohemia and Moravia; foundation of the independent Slovak state September 1939: beginning of the Second World War	1935: first female university professor at Charles University—Milada Paulová in the field of history of Slavic nations and Byzantine studies
May 1945: end of the Second World War in Europe; rebuilding of the Czechoslovak Republic; Edvard Beneš becomes the president	1945: preparatory committee of the Council of Czechoslovak Women led by Milada Horáková; regulation on equal pay of women
February 1948: communist coup in Czechoslovakia; Klement Gottwald becomes the president	1948: first woman in Czechoslovak government—Ludmila Jankovcová
1949–1954: political trials in Czechoslovakia 1950: trial and execution of Milada Horáková 1949: foundation of the Council for Mutual Economic Assistance 1955: foundation of the Warsaw Pact	1949: first female rector—Jiřina Popelová, Palacký University in Olomouc
1968: Warsaw Pact invasion of Czechoslovakia; end of the democratisation process and liberalisation in Czechoslovakia 1970: beginning of 'normalisation' 1977: Charter 77 1986: Mikhail Gorbachev becomes the leader of the USSR– beginning of changes in Eastern Europe November 1989: Velvet Revolution in Czechoslovakia; return of capitalist economy 1992: disintegration of Czechoslovakia 1 Jan 1993: foundation of the Czech Republic	

Index

Z. Jastrzembská et al., *Czech Women Philosophers and Scientists*,
SpringerBriefs in History of Science and Technology,
https://doi.org/10.1007/978-3-030-61516-1

The manufacturer's authorised representative in the EU is Springer Nature Customer Service Centre GmbH, Europaplatz 3, 69115 Heidelberg, Germany. If you have any concerns regarding our products, please contact ProductSafety@springernature.com

Printed and bound by CPI Group (UK) Ltd, Croydon, CR0 4YY
22/09/2026
02221064-0016